RAND NATIONAL SECURITY RESEARCH DIVISION

T0122812

Supporting Employers in the Reserve Operational Forces Era

Are Changes Needed to Reservists' Employment Rights Legislation, Policies, or Programs?

Susan M. Gates, Geoffrey McGovern, Ivan Waggoner, John D. Winkler, Ashley Pierson, Lauren Andrews, Peter Buryk

Prepared for the Office of the Secretary of Defense
Approved for public release; distribution unlimited

The research described in this report was prepared for the Office of the Secretary of Defense (OSD). The research was conducted within the Forces and Resources Policy Center of the RAND National Defense Research Institute, a federally funded research and development center sponsored by the Office of the Secretary of Defense, the Joint Staff, the Unified Combatant Commands, the Navy, the Marine Corps, the defense agencies, and the defense Intelligence Community under Contract W74V8H-06-C-0002.

Library of Congress Control Number: 2013945722

ISBN: 978-0-8330-8091-2

The RAND Corporation is a nonprofit institution that helps improve policy and decisionmaking through research and analysis. RAND's publications do not necessarily reflect the opinions of its research clients and sponsors.

Support RAND—make a tax-deductible charitable contribution at www.rand.org/giving/contribute.html

RAND® is a registered trademark.

Preface

Employer Support of the Guard and Reserve (ESGR), a U.S. Department of Defense (DoD) office, asked the RAND National Defense Research Institute to study the effects that using the Reserve Components (RCs) as an operational force can have on employers. The primary purpose of the study was to consider whether changes are needed to the Uniformed Services Employment and Reemployment Rights Act (USERRA) (Pub. L. 103-353, 1994), ESGR support programs, or RC activation and deployment policies given the increased mobilization of the National Guard and Reserve and the continuing need to balance the rights, duties, and obligations of employers, RC members, and RC members' families. The study involved the review and analysis of existing research and data related to USERRA and the effects on employers of employee absences more generally, an analysis of the DoD National Survey of Employers fielded in 2011, focus groups with employers conducted in 2012, interviews with RC chiefs conducted in 2011, and a legal and legislative history review of USERRA. This report describes key findings from the analysis.

This report will be of interest to policymakers in DoD and the U.S. Department of Labor, human resource officials in government agencies and other organizations, military commanders, and RC members. This research was sponsored by ESGR and conducted within the Forces and Resources Policy Center of the RAND National Defense Research Institute, a federally funded research and development center sponsored by the Office of the Secretary of Defense, the Joint Staff, the Unified Combatant Commands, the Navy, the Marine Corps, the defense agencies, and the defense Intelligence Community.

For more information on the RAND Forces and Resources Policy Center, see http://www.rand.org/nsrd/ndri/centers/frp.html or contact the director (contact information is provided on the web page).

Contents

Figures

Tables

Summary

The Uniformed Services Employment and Reemployment Rights Act (USERRA) was designed to prevent hiring discrimination and bolster job protection for members of the armed forces, including those of the Reserve Components (RCs).[1] Under USERRA, it is against the law for an employer to refuse to hire a reservist, guardsman, veteran, or someone applying to enter the military services on the basis of the applicant's military service affiliation. Moreover, service members are guaranteed reemployment by their former employers after periods of military duty. Since 1994, this landmark employment rights legislation has provided assurance to men and women weighing the decision to volunteer to defend the United States and is a tangible expression of gratitude for their immense sacrifice.

Although Congress considered the potential costs that USERRA might impose on employers in crafting the original legislation, the way in which the U.S. Department of Defense (DoD) uses the RCs has changed dramatically since 1994—potentially altering the cost-benefit calculation on which the passage of USERRA was based. During the Cold War, the RCs were largely viewed as a strategic reserve that could be called upon in the event of a major contingency operation. RC members participated in regular drill activities (one weekend a month plus an additional two-week period each year) but did not view extended activation as a likely event.[2] This model of the RCs began to change with the first Gulf War in 1991, and then changed dramatically after September 11, 2001.

Measured in terms of the number of duty days for members of the RCs, the aggregate burden that employers are asked to shoulder in support of military operations more than doubled between fiscal year (FY) 1993 and FY 1996—from 5.3 million duty days to a sustained level of 12 million to 13 million duty days. After

---

[1] The term *RC* is used to refer to all seven Guard and Reserve organizations: Army National Guard, Army Reserve, Navy Reserve, Marine Corps Reserve, Air National Guard, Air Force Reserve, and the Coast Guard Reserve (10 U.S.C. §101).

[2] An activation is an order to active duty for purposes other than training. An activation may be voluntary or involuntary. An activation may or may not be associated with a deployment. A deployment involves the movement of forces for the purposes of a military operation.

2001, during the global war on terrorism, the burden continued to grow, reaching 68.3 million duty days in 2005. By FY 2010, the number of duty days had declined to 37.5 million. Thus, although it is substantially lower than the FY 2005 peak, the number of duty days in FY 2010 still far exceeded the norm at the time USERRA was passed.

Activation and deployment are now common occurrences for RC members: Approximately one-third of RC members surveyed in January 2011 reported that they had been activated in the previous two years, with more than three-quarters of those experiencing deployment.[3]

The shift from a strategic to an operational reserve implies that employers of RC members are more likely to experience a duty-related employee absence than they would have been when USERRA was passed. Survey data reveal that those absences typically exceed 30 days.

Given the increased mobilization of the National Guard and Reserve and the continuing need to balance the rights, duties, and obligations of employers, RC members, and RC members' families, DoD asked the RAND National Defense Research Institute to consider whether changes are needed to USERRA, Employer Support of the Guard and Reserve (ESGR) programs, or RC activation and deployment policies. To address this policy question, the study focused on the following research questions:

- What are the legal protections provided by USERRA, what obligations do they impose on employers, and what are the areas of ambiguity?
- To what extent do employers understand their obligations under USERRA?
- What factors influence the cost of USERRA protections for employers?
- What changes to USERRA or to DoD policy would employers find useful in fulfilling their obligations under USERRA?

To address these questions, we undertook five key research tasks: a review of prior literature and existing data, descriptive and multivariate analysis of data collected by DoD in 2011 through the National Survey of Employers and between 2007 and 2010 through the Status of Forces Surveys of Reserve Component Members (SOFS-Rs), legislative and legal historical review of USERRA, interviews with each of the RC chiefs conducted in 2011, and interviews with 16 employers conducted in 2012. Table S.1 describes how the research tasks informed our answers to the research questions.

This report presents findings from this multimethod approach. The literature review, analysis of data from the SOFS-Rs, and interviews with RC chiefs also serve to provide important contextual information that aids in the interpretation of the research findings. We synthesize the findings from the research questions described

---

[3] For National Guard members, these duty figures would include duty in support of state missions. USERRA covers duty in support of federal missions. National Guard duty in support of state missions is covered by state laws.

Table S.1
Information Sources Used to Address Research Questions

Research Question	Information Source
What are the legal protections provided by USERRA, what obligations do they impose on employers, and what are the areas of ambiguity?	Historical review of USERRA, DoD National Survey of Employers, interviews with employers
To what extent do employers understand their obligations under USERRA?	DoD National Survey of Employers, interviews with employers and with RC chiefs, SOFS-Rs
What factors influence the cost of USERRA protections for employers?	Review of prior research, DoD National Survey of Employers, interviews with employers and with RC chiefs, SOFS-Rs
What changes to USERRA or to DoD policy would employers find useful in fulfilling their obligations under USERRA?	DoD National Survey of Employers, interviews with employers and with RC chiefs

here to develop recommendations regarding potential changes to USERRA, ESGR programs and RC activation, and deployment policies.

USERRA Provides Protection Against Discrimination and Provides Reemployment Rights

USERRA was written to encourage service in the U.S. armed forces by removing some barriers to military service that are related to civilian employment. In its current form, there are two main aspects of the law: (1) general antidiscrimination protection based on veteran status and (2) reemployment rights after periods of military duty. Both categories of protection offer valuable benefits to veterans, as well as to civilians considering joining the military. As written, the law is strongly pro–service member. The law explicitly protects both voluntary and involuntary duty and, unlike other employment laws, does not exempt employers from its provisions based on their size. However, the law does provide for exceptions to the reemployment provisions—for example, in instances in which the provisions might cause the employer undue hardship.

Employer Impact of USERRA Stems from Duty-Related Absences

Our framework for understanding the impact of duty-related absences in the context of USERRA builds on two key insights in the prior literature. First, the net effect for employers stems from both direct and indirect impacts of dealing with the absence in a way that complies with USERRA. Direct impacts are those associated with recruiting, screening, hiring, and training replacement workers, as well as retraining the reservist upon his or her return from military duty (as needed). Direct impacts also include the relative cost of the replacement worker (which may be positive or negative, depending on the cost of the absent reservist) and the cost of providing benefits to the reservist during his or her absence (either voluntarily or as required by law). In addition, indirect effects, such as lost business, productivity, or opportunity for growth, may be incurred. Second, the impact will vary across employers and will also be influenced by who is getting deployed and for how long.

Figure S.1 describes our conceptual framework. By establishing employment and reemployment rights, USERRA has the potential to impose costs on employers. Those costs are influenced by characteristics of the absence, the employee who is activated and his or her position with the employer, the employer, and the overall economy. DoD policy, as well as DoD and ESGR programs, targeting employers can influence costs *indirectly* through their effect on the nature of absences, the characteristics of employees who are activated and their positions with their employers, and potentially the characteristics of employers. DoD and ESGR programs can also have a *direct* influence on some of those costs. A key focus of our analysis was to identify any systematic relationships between characteristics of the absence, position or employee, or employer on the one hand and the cost of absences to the employer on the other and to consider what types of policies might provide an effective response.[4]

Findings

Legal Protections Provided by USERRA Are Generally Clear and Consistent with Other Employment Laws

We considered whether there is sufficient need for a rewrite of the USERRA legislation based on USERRA's legal and legislative history. Whereas USERRA's predecessor laws (Selective Training and Service Act [Pub. L. 76-783, 1940], and the Vietnam Era Veterans' Readjustment Assistance Act, commonly known as the Veterans' Reemployment Rights Act [Pub. L. 92-540, 1972]) were deficient in terms of content, DoD policy interface, and general harmony with employment protections related to things other than military service or affiliation, USERRA is generally clear and consistent with other employment laws. As such, we conclude that there is no need for substantial revision to the legislation. Many of the challenges employers face stem from RC activation rates and utilization policy rather than from the employment and reemployment protections of USERRA itself.

---

[4] Because much of the data used in this study were collected during a significant economic downturn, we are not able to identify relationships between characteristics of the economy and the cost of absences, although such relationships have been suggested in the literature.

Figure S.1
Conceptual Framework Linking a Duty-Related Absence to Employer Impact

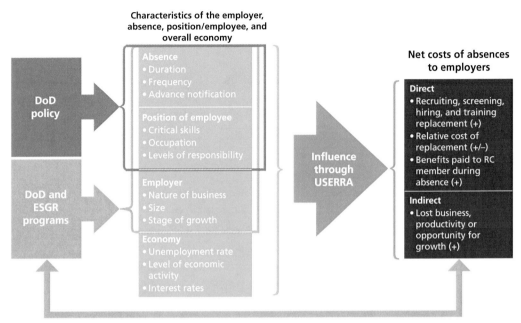

RAND *RR152-S.1*

Understanding of USERRA, Employer Obligations, and Where to Go for Help Is Incomplete

Our analysis reveals that employer knowledge of USERRA and ESGR is incomplete. RC employers express greater familiarity with USERRA and confidence in their ability to comply with the law and awareness of ESGR than employers that do not employ an RC member. Nevertheless, roughly one-quarter of RC employers disagreed with the statement that they knew all they needed to know in order to remain in compliance with USERRA. Small employers were more likely to disagree with this statement than large employers. Well below half of RC employers were aware of the Statement of Support and programs and awards sponsored by ESGR. Many employers expressed an interest in the types of programs, awards, and supports already offered by ESGR.

Low response rates to the survey and employers' unwillingness to participate in interviews on this topic suggest that USERRA is not a high-priority issue for most employers except when they are dealing with duty-related absences.

In addition, there is extant policy that instructs the RCs to identify a representative who can consider requests from civilian employers for adjustments to reserve duties in cases in which such duties would have adverse effects on an employer's businesses (Under Secretary of Defense for Personnel and Readiness, 1997). Currently under review by DoD, this instruction, DoD Instruction 1205.12 has the potential to meet some of the concerns raised by employers that do face hardships to their business activities. Our study suggests that knowledge of such policy among employers is low. Details about the policy and its implementation are not readily available, suggesting the need for further DoD review.

The Impact That USERRA Protections Can Have on Employers Is Highly Varied and Influenced by a Wide Range of Factors

Duty-related absences can impose a substantial burden on employers in some circumstances, but the impacts are highly varied and influenced by a wide range of factors. The impact of USERRA protections stems from the direct and indirect implications of duty-related absence for business operations, rather than from the costs of the specific employment and reemployment rights contained in the legislation. These implications of absences include the need to manage workload, the need to replace the RC member, loss of business, costs of health and other benefits, and lost productivity. Among employers that experience a business impact, issues related to managing workload are most common. However, each type of impact was reported by at least one employer, and small employers were more likely than large employers to report a loss of business.

Our study confirms that factors suggested by prior research and described in our conceptual framework contribute to employer impact. We find evidence that the following factors are associated with greater employer impact:

- longer or more-frequent absences
- the absence of highly skilled employees, key personnel, or employees who are "one of a kind" in their organization
- employer characteristics, including employer size, with smaller employers being more likely to experience impacts.

However, our analysis does not identify any single factor or combination of factors as clear drivers of impact. For example, although employers that experience longer absences are more likely to report an impact, many employers that experienced absences longer than a year reported no impact.

Instead, our analysis suggests that it is a more idiosyncratic combination of factors, including employer circumstances at the time of activation, that result in a significant employer impact. School districts, businesses that depend on personal relationships with clients or customers, and businesses that rely on highly skilled or trained employees are more likely to experience problems and may need more advance notice to plan for the absence. In addition, employers reported challenges in dealing with the activation of the only employee of a certain type in the organization—whether or not that individual is highly skilled. Because it is not possible to shift the work to another employee, more advance planning is needed in these situations.

Employers reported a need to plan for the return of RC members, as well as for their pending absences. Some employers indicated that planning for the return is inhibited by a lack of information about whether the RC members will return to work after duty and when the duty will end. For a highly skilled employee, an employer can face a need to retrain the employee when he or she returns, particularly in the case of a long absence. Some employers mentioned challenges involved in terminating replacement workers, including the resulting increase in the costs of unemployment insurance. In addition, about 15 percent of employers that experienced a duty-related absence reported that the RC members faced problems, such as difficulty interacting with clients or emotional problems, upon return to civilian employment.

No Single Change to DoD Policy or Support Programs Would Address the Concerns of All Employers

There was no consensus among employers as to the support programs that would be most useful to their businesses in the event of a duty-related absence. The largest share of employers reported that no support programs were needed, but each of the survey options was selected by at least one employer. Employers did express a desire for advance notification of both absence and return, and many employers wanted clear and consistent documentation of the term of duty directly from DoD. Some employers—most notably, school systems, employers with seasonal or cyclical work schedules, smaller employers, and employers facing the absences of key personnel or multiple personnel (including first responders)—requested more flexibility and employer input in scheduling both training duty and deployment. Other common suggestions provided by employers were limits to the duration or frequency of voluntary duty, replacement assistance (possibly at government expense), government contracting preferences for RC employers, and means for remaining in contact with RC members during their terms of duty.

Most employers did not state a preference for or against shorter, more-frequent duty versus longer, less-frequent duty. And although two-thirds of all employers expressed a preference for combining training and deployment in the event of a long deployment-related activation, one-third of employers preferred to have a gap between training and deployment.

Federal Agencies Employ a Disproportionate Number of Reserve Component Members and Appear to Value and Support Those Members as Much as Other Employers Do

The federal government employs a disproportionately large number of RC members. Whereas the federal government employs less than 2 percent of the U.S. workforce (Falk, 2012), the 2011 RC member survey reveals that it employs 19 percent of the RC members who are employed full time. RC members employed by the federal government are neither more nor less likely to experience USERRA problems and are neither more nor less likely to describe their employers as "supportive" or "very supportive" of their military duty compared with private, for-profit employers. Compared with private, for-profit employers, federal employers expressed a higher level of awareness of USERRA obligations and were more likely to view RC members more favorably than employees who are not RC members. However, the percentage of federal employers reporting that RC members returning from activation had problems with substance abuse, a service-related disability, or increased stress was more than twice as high as for the overall National Survey of Employers population.

Recommendations

We conclude that **there is no need for substantial revision to the legislation**. Our findings suggest that DoD can most effectively support employers through modifications to DoD programs and policies.

Providing Greater Flexibility to Employers Holds More Promise Than Broad Changes to USERRA or Utilization Policy

Our findings suggest that across-the-board changes to USERRA, utilization policy, or duty structure applied to all RC members and all employers will not resolve employer impacts uniformly, given the sporadic and varied nature of problems faced by employers.

We recommend that DoD consider implementing a more extensive appeal process that would not be limited to specific categories of employers but rather could be accessed by employers on an as-needed basis. Such an appeal process, similar to that used in the United Kingdom, would allow DoD to effectively address the most-significant instances of employer impact and balance its needs with those of the employer. DoD already allows individual RC members to petition for relief from activation. Extant policy also provides employers with access to a similar process. Strengthening such policy, in terms of both implementation and promoting greater knowledge in the employer community, could ameliorate the most-significant instances of employer impact.

DoD Should Explore Ways to Standardize and Expand Communication with Employers About Duty-Related Absences and Available Resources

Since 2004, RC members have been required to provide DoD with information on their civilian employers and job skills and to notify DoD of any job changes (DoD, 2004). The data provided by RC members are retained in the Civilian Employment Information (CEI) database. Many employers would like to see improvements to communication from DoD about duty-related absences—particularly duty start and end dates. DoD should consider developing and preparing a simple, easy-to-read form for employers that documents duty start and end dates. The form could include information on the time frame in which the RC member must notify the employer of his or her intent to return to employment per USERRA provisions. In addition, DoD should provide employers with updated information whenever duty is extended for the benefit of those employers that are not receiving updates directly from the RC members. Finally, DoD could use these notifications as an opportunity to provide a USERRA information packet and ESGR contact information to employers at the start of duty and information on DoD resources for service members and veterans at the end of duty. Changes to the CEI program may be required to effectively support such improvements.

As part of these efforts, DoD should ensure that the resources available to help employers and RC members deal with common issues that arise in the transition back to civilian life after a deployment are appropriate and adequate. DoD should also ensure that RC employers (as well as RC members) are aware of these resources. Most employers value the RC members they employ and are supportive of their service to the United States. In cases in which an RC member returns from a deployment with problems, employers want to help. DoD should improve access to information on support programs that are available to RC members. DoD should also explore why federal employers are more likely than private, for-profit employers to report that employees are returning from duty with employment-related challenges. It may be that federal employers have a higher level of awareness of possible issues or that RC members who are federal employees are experiencing such challenges at a higher rate.

Of course, these suggestions will be useful to employers only to the extent that DoD maintains up-to-date information about employers of RC members. Responses to the National Survey of Employers suggest that continued efforts to improve this information are needed.

ESGR Should Continue Its Efforts to Promote Awareness of USERRA

Our findings suggest that ESGR should continue its efforts to promote awareness of USERRA and other employer-related programs and policies—especially among known employers of RC members. ESGR should consider partnering with the U.S. Department of Labor and the U.S. Small Business Administration to improve the dissemination of information about USERRA, particularly to smaller employers.

Employer Peer Resource Networks Could Assist Employers in Dealing with Duty-Related Absences

Employers that have had successful experiences with duty-related absences can be a useful resource for other companies facing such absences for the first time. A network of employers of different sizes and in different industries could be a powerful peer resource for other employers. DoD should consider soliciting employers to serve as peer mentors and developing an award program to recognize employers that provide this valuable service. Such a program would provide another avenue for engaging RC employers while providing them with direct support.

Acknowledgments

We are indebted to the thousands of employers that took the time to respond to surveys or participate in interviews about the implications that duty-related absences have on their businesses.

We are grateful to our peer reviewers, Ellen M. Pint and Zoe Morrison, who completed a careful review of an earlier draft of this document and contributed extensive suggestions for improving the report. The final report is much improved thanks to their input.

Many former and current ESGR staff members provided extensive guidance and assistance throughout the course of this project. We are grateful for the support and access to information, input, and comments on a prior draft of this report provided to us by Curtis Bell, Raymond Carney, Jeff Graber, Virginia Hyland, Ted Fessel, Boyd Parker, Paul Vining, Samantha Walker, Ron White, and Ronald G. Young.

Kimberly Hylton, Fawzi Al Nassir, Karen Wessels, and Kimberly Williams from the Defense Manpower Data Center provided us with data from the DoD National Survey of Employers, answered numerous questions about the data, provided us with data from which we could sample employers for our interviews, and provided input on our Office of Management and Budget (OMB) packet for the employer interviews.

We are also grateful to Patricia Toppings (Washington Headquarters Service), Jill Merkley (U.S. Navy), and Caroline Miner (Office of the Secretary of Defense), who helped us navigate the OMB clearance process.

We would like to thank Nono Ayivi-Guedehoussou, Neema Iyer, Jake Solomon, and Donna White, all from RAND, for their assistance in contacting and screening employers regarding participation in this study.

Thomas Liuzzo, Daniel Wolf, and Patrick Zimmerman of the Office of the Assistant Secretary of Defense for Reserve Affairs graciously provided us with information about the utilization of RC members and requests from employers for duty deferrals that we reference in this report.

Allison Elder of RAND provided guidance and input on the employer perspective at an early stage of this project. Suzy Adler assisted with data acquisition. Donna White provided critical help with document formatting. Philip Hall-Partyka compiled information on the margin of error for survey tabulations. Susan Paddock provided statistical consulting on our analytical methods, advised on the appropriate interpretation of model parameters, and reviewed some of our programming code. Barbara Bicksler provided suggestions for improving the document. Christina Steiner assisted with information gathering and literature review. Lisa Bernard carefully edited the final copy.

We take full responsibility for any remaining errors.

Abbreviations

AFR	Air Force Reserve
AG	U.S. Attorney General
ANG	Air National Guard
AO	adjudication officer
ARNG	Army National Guard
CBO	Congressional Budget Office
CEI	Civilian Employment Information
CIA	Central Intelligence Agency
DMDC	Defense Manpower Data Center
DoD	U.S. Department of Defense
DoDI	Department of Defense instruction
DOJ	U.S. Department of Justice
DOL	U.S. Department of Labor
DUNS	Data Universal Numbering System
EN	Employer Notification
EP	Employer Partnership of the Armed Services
EPF	Employment Policy Foundation
ESGR	Employer Support of the Guard and Reserve
FBI	Federal Bureau of Investigation
FMLA	Family and Medical Leave Act

FR	first responder
FY	fiscal year
GAO	U.S. Government Accountability Office
MOD	UK Ministry of Defence
MSPB	Merit Systems Protection Board
OASD(RA)	Office of the Assistant Secretary of Defense for Reserve Affairs
OEF	Operation Enduring Freedom
OIF	Operation Iraqi Freedom
OMB	Office of Management and Budget
OPM	Office of Personnel Management
OSC	U.S. Office of Special Counsel
PaYS	Partnership for Youth Success
PSM	program support manager
PTSD	posttraumatic stress disorder
RC	Reserve Component
SBA	U.S. Small Business Administration
SOFS-R	Status of Forces Survey of Reserve Component Members
SOS	Statement of Support
TBI	traumatic brain injury
USAR	U.S. Army Reserve
USERRA	Uniformed Services Employment and Reemployment Rights Act
VETS	Veterans' Employment and Training Service

CHAPTER ONE

Introduction

The Uniformed Services Employment and Reemployment Rights Act (USERRA) was designed to prevent hiring discrimination and bolster job protection for members of the armed forces, including those of the Reserve Components (RCs). Under USERRA, it is against the law for an employer to refuse, on the basis of military status, to hire a veteran, an RC member, or someone in the process of enlisting. Moreover, after a period of military service, service members are guaranteed reemployment by their civilian employers when their duty is over. USERRA's reemployment protections are particularly relevant for RC members because their participation in the military typically coincides with civilian employment. Since 1994, this landmark employment rights legislation has provided assurance to men and women weighing the decision to volunteer to defend the United States and is a tangible expression of gratitude for their immense sacrifice.

The world has changed since USERRA was instituted, however. For one thing, the armed forces have become much more reliant on the RCs, effectively redesignating them from a strategic to an operational reserve in 2008 (U.S. Department of Defense [DoD] Directive 1200.17, 2008). Twenty years ago, an RC member could reasonably expect a one-time occasion of involuntary mobilization. Employers, in turn, could expect their citizen-soldier employees to have predictable absences related to weekend and two-week annual training and limited time away from the job in the case of war or national emergency. Now, RC members are called to serve much more frequently. They participate in longer tours of duty, conducting operations in Iraq and Afghanistan, as well as supporting missions in the continental United States. Although employers have played a critical role in helping RC service members carry out their missions, employers no longer know what to expect of RC members' schedules. This changed operational climate has led to a difficult question: Are USERRA's protections too burdensome for employers?

Background

USERRA is an employment law whose primary objective is to grant specified rights to employees who are or were members of the armed services. The legislation strives both to prevent employers from discriminating against RC members in the hiring process and to provide RC members with benefits and protections against discrimination based on RC status and duties once hired. These aims are similar to those of other employment laws that provide benefits and protections to other populations. These include the Civil Rights Act (Pub. L. 88-352, 1964), the Age Discrimination in Employment Act (Pub. L. 90-202, 1967), amendments to Title VII of the Civil Rights Act of 1964 to prohibit sex discrimination on the basis of pregnancy (Pub. L. 95-555, 1978), the Americans with Disabilities Act of 1990 (Pub. L. 101-336, 1990), and the Family and Medical Leave Act (Pub. L. 103-3, 1993) (FMLA).

As discussed in Dixon et al. (2007, pp. 41–44), employment laws and regulations originate from a perceived need or desire on the part of policymakers to protect or benefit a class of citizens. In conveying employment protections and benefits, policymakers recognize that they may impose a cost on employers:

> [I]n striving to strike a balance between costs and benefits, policymakers adjust the application or enforcement of employment-related regulations according to firm size due to the belief that a given regulation or regulatory policy will impose a greater relative cost on a smaller firm. (Dixon et al., 2007, p. 41)

For example, the Fair Labor Standards Act (Pub. L. 75-718, 1938), which establishes the federal minimum wage, overtime provisions, and restrictions on the use of child labor, applies to businesses with two or more employees. The FMLA, which requires employers to allow employees 12 weeks of unpaid leave for specific reasons, such as providing care for a seriously ill family member, applies only to businesses with 50 or more employees. In this context, it is notable that USERRA applies to all businesses, regardless of the number of employees.

As we make clear in Chapter Three of this report, in passing USERRA in 1994, Congress was well aware that the law would impose costs on employers. Nevertheless, Congress took the view that protecting RC members against discrimination and ensuring they would have a job after returning from an absence due to military duty was not only the right thing to do but also critically important to ensuring a supply of people willing to participate in the RCs.

The Shift from a Strategic Reserve to an Operational Reserve

Although Congress considered the potential costs USERRA might impose on employers in crafting the original legislation, DoD's use of the RCs has changed dramatically since 1994—potentially altering the cost-benefit calculation on which the passage of USERRA was based. During the Cold War, the RCs were largely viewed as a strategic reserve that could be called upon in the event of a major contingency operation. RC members participated in regular drill activities (one weekend per month plus an additional two-week period each year) but did not view extended activation as a likely event. This model of the RC began to change with the first Gulf War in 1991, and then changed dramatically after September 11, 2001.

Figure 1.1 presents the number of RC duty days in support of federal missions from fiscal year (FY) 1986 to FY 2010. The number of duty days for members of the RCs can be viewed as a rough reflection of the aggregate burden that private-sector employers are asked to shoulder in support of military operations. As illustrated by Figure 1.1, the size of that burden more than doubled between FY 1993 and FY 1996— from 5.3 million duty days to a sustained level of 12 million to 13 million duty days (see also Buck et al., 2008, Figure 1, p. 6). After 2001, during the global war on terrorism, the burden grew by more than a factor of ten, reaching 68.3 million duty days in 2005. By FY 2010, the number of duty days had declined to 37.2 million. Although it is substantially lower than the FY 2005 peak, the number of duty days in FY 2010 far exceeds the norm at the time USERRA was passed.

Activation and deployment are now common occurrences for RC members: Approximately one-third of RC members surveyed in January 2011 reported that they had been activated in the preceding two years, with more than three-quarters of those

Figure 1.1
Annual Number of Active-Duty Days for Reserve Component Members, by Fiscal Year

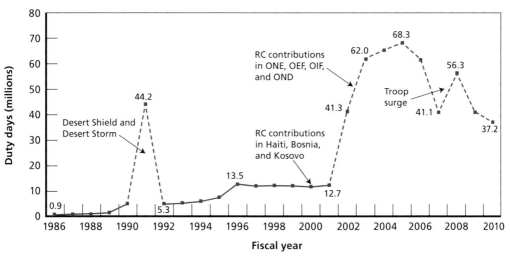

SOURCE: Office of the Assistant Secretary of Defense for Reserve Affairs.
NOTE: ONE = Operation Noble Eagle. OEF = Operation Enduring Freedom. OIF = Operation Iraqi Freedom. OND = Operation New Dawn.
RAND RR152-1.1

experiencing deployment. Our analysis of data from a DoD survey of RC members conducted in January 2011 found that 34 percent of respondents reported having been activated in the 24 months prior to the survey. Of those activated, 93.6 percent reported that their activations were longer than 30 days. Of those activated for longer than 30 days, 79 percent reported being deployed (DoD, 2011). Similar rates of activation were reported in Status of Forces Surveys of Reserve Component Members (SOFS-Rs) going back to 2008.

The shift from a strategic to an operational reserve implies that employers of RC members are more likely to experience a duty-related employee absence and that the duration of the absence is likely to be longer than it would have been when USERRA was passed. SOFS-R data reveal that those absences typically exceed 30 days.

The shift in DoD policy regarding the utilization of RC members since 1994 has likely led to an overall increase in the burden military absences place on employers, as well as changes in the way in which that burden is distributed across employers. Some have argued that this increased burden has reduced employer willingness to hire military veterans (Harrell and Berglass, 2012).

Approach

Employer Support of the Guard and Reserve (ESGR), a DoD office, asked the RAND National Defense Research Institute to study the effects that using the RCs as an operational force can have on employers. The primary purpose of the study is to consider whether changes are needed to USERRA, ESGR programs, or RC activation and deployment policies given the increased mobilization of the National Guard and Reserve and the continuing need to balance the rights, duties, and obligations of employers, RC members, and RC members' families. To address this policy question, the study focused on the following research questions:

- What are the legal protections provided by USERRA, what obligations do they impose on employers, and what are the areas of ambiguity?
- To what extent do employers understand their obligations under USERRA?
- What factors influence the cost of USERRA protections for employers?
- What changes to USERRA or to DoD policy would employers find useful in fulfilling their obligations under USERRA?

To address these questions, we undertook five key research tasks: a review of prior literature and existing data, descriptive and multivariate analysis of data collected by DoD through the National Survey of Employers and through the SOFS-R, legislative and legal historical review of USERRA, interviews with each of the RC chiefs, and interviews with 16 employers conducted by RAND. Additional information about each research task is provided in the sections that follow. Table 1.1 describes how the research tasks informed our answers to the research questions.

This report presents findings from this multimethod approach. The literature review, analysis of data from the SOFS-Rs, and interviews with RC chiefs also serve to provide important contextual information that aids in the interpretation of the research findings. We synthesize the findings from the research questions to develop recommendations regarding potential changes to USERRA, ESGR programs, and RC activation and deployment policies.

Review of Prior Literature and Existing Data

In conducting this study, we reviewed prior research on USERRA, the cost of military absences for employers, and trends in DoD's utilization of RC members. We also reviewed the literature on the cost that non–military-related absences of employees can have for employers. Although our literature review focused on peer-reviewed publications, we also reviewed reports and articles targeted to human resource professionals—not all of which were peer reviewed. We also reviewed studies published by the U.S. Department of Labor (DOL) and by ESGR on USERRA complaints and violations and their resolutions.

Legal and Legislative History of the Uniformed Services Employment and Reemployment Rights Act

Our study involved a thorough documentation of the provisions of USERRA, their historical context, and the manner in which those provisions have been amended and interpreted by the courts. USERRA was signed into law on October 12, 1994, and has been revised multiple times since then (Pub. L. 104-275, 1996; Pub. L. 105-368, 1998; Pub. L. 106-419, 2000; Pub. L. 108-454, 2004; Pub. L. 110-389, 2008; Pub. L. 112-56,

Table 1.1
Information Sources Used to Address Research Questions

Research Question	Information Sources
What are the legal protections provided by USERRA, what obligations do they impose on employers, and what are the areas of ambiguity?	Historical review of USERRA, DoD National Survey of Employers, interviews with employers
To what extent do employers understand their obligations under USERRA?	DoD National Survey of Employers, interviews with employers and with RC chiefs, SOFS-Rs
What factors influence the cost of USERRA protections for employers?	Review of prior research, DoD National Survey of Employers, interviews with employers and with RC chiefs, SOFS-Rs
What changes to USERRA or to DoD policy would employers find useful in fulfilling their obligations under USERRA?	DoD National Survey of Employers, interviews with employers and with RC chiefs

2011). Here, we highlight key provisions of the law as of January 2012. Further details are provided in Appendix A, published separately online.

We also review and document USERRA's legislative history, the USERRA claim process, and the legal history. Our review traces the major pieces of legislation leading up to USERRA and highlights why, at each stage, new legislation was required to meet the changing needs of the country. We identify early policy debates that gave rise to USERRA in order to reveal the underlying challenges the legislation was designed to remedy.

Our approach to reviewing the history of legal challenges to USERRA focuses almost exclusively on appellate courts. We focus on appellate court opinions because they resolve uncertainties found within the law itself; in contrast, the responsibility of district courts is to determine the facts or the events that took place in a specific case. Because our primary interest lies with the policy issues surrounding the law, and not the details of individual cases that are brought before the court, an appellate review is the more appropriate approach.

Analysis of Previously Collected Data

We analyzed survey data collected by DoD through a nationally representative sample of employers and of RC members.[1] In this report, we present findings from these analyses in an integrated and topical manner—providing both the employer and the RC member perspectives on an issue (as appropriate). Much of our analysis involved descriptive tabulations, often of subsets of the overall survey sample. We also conducted factor analysis and regression analysis of the survey data to explore relationships between survey responses and employer characteristics. Below, we provide a description of the data sets and our analytical approach.

We also reviewed selected reports from DOL on the incidence of new cases opened by the Veterans' Employment and Training Service (VETS), as well as data from ESGR on the number of calls received concerning USERRA issues. These data sources provided background and context but were not comprehensive enough for statistical analysis.

2011 U.S. Department of Defense National Survey of Employers

In 2011, the Defense Manpower Data Center (DMDC) surveyed business establishments (including private-sector employers, nonprofit organizations, and government offices) across the country, sampling both employers that had employed RC members since June 2007 and employers that had not employed RC members during that time. Of a sample of nearly 80,000 employers, approximately 10,500 employers returned completed surveys, for a response rate of 17 percent for RC employers and 5 percent for non-RC employers (that is, employers not known to have recently employed RC members). The RC employers were selected from the Civilian Employment Informa-

---

[1]   RAND was not involved in the survey design or data collection.

tion (CEI) system maintained by DMDC; the stratified random sample of non-RC employers was pulled from the Dun and Bradstreet database and excluded employers in the CEI system.

The sample was representative of all businesses in the United States with paid employees and thus excluded the self-employed. Each location of an employer with more than one office or location was eligible for inclusion. Stratification was by state and RC employer status.

The survey asked each respondent about his or her experiences with RC members, including experiences related to military absences, the ways in which a military absence affects the business, how the employer deals with an RC member's return to work, potential measures to improve DoD's relationship with employers, knowledge of USERRA and employer support programs, and general employment practices related to RC members.

Our analysis of the data from this DoD National Survey of Employers is presented in Appendix B, published separately online. This report highlights key findings most relevant to the research questions. In analyzing the data, we focused on exploring variation in responses by employer characteristic. We also reviewed the responses to open-ended questions in the survey. For example, each respondent was asked to write in suggestions for additional measures that would be helpful for his or her business in the event of the activation of an RC employee. (The survey options were hiring incentives, replacement assistance, opportunity to reschedule military duty, and reimbursement of employer expenses.) DMDC provided these responses to us separately from the main survey data, and the responses were associated only with information on employer size and whether the respondent was drawn from the CEI or non-CEI sample. As a result, our ability to relate these comments to employer characteristics is limited. With the exception of the analysis of open-ended survey comments, all survey results presented in this report are weighted and thus representative of the population of U.S. businesses or the relevant subsample of U.S. businesses. Our analysis was performed using Stata® (StataCorp, 2007).

Status of Forces Surveys of Reserve Component Members

We analyzed data from the SOFS-Rs conducted between June 2008 and January 2011. Appendix C, published separately online, presents findings from the analysis of the January 2011 survey of 120,724 RC members that resulted in 21,873 valid responses. This report highlights key findings most relevant to the research questions. The SOFS-Rs are administered to members of the Army National Guard (ARNG), U.S. Army Reserve (USAR), U.S. Navy Reserve, U.S. Marine Corps Reserve, Air National Guard (ANG), and Air Force Reserve (AFR) who have at least six months of service. The survey did not include the most-senior (flag-rank) officers. With use of weights, the results of these surveys are representative of the population; all results presented here are weighted and thus represent the population of RC members.[2] Following DMDC Report 2009-008 (DMDC, 2009), this analysis excludes those respondents who stated that they were not in an RC at the time of the survey.

The SOFS-Rs cover a range of topics related to the relationship between an RC member and his or her employer prior to, during, and after a duty-related absence, including questions about whether the RC member returned to work for the same employer and whether he or she experienced any USERRA-related issues. The survey also obtained information about the characteristics of the employers. To the extent possible, our analysis explored the relationship between employer characteristics and survey responses. Our analysis was performed using Stata (StataCorp, 2007).

Focus Group and Individual Interviews of Employers

We conducted interviews with 16 employers in focus groups and individually via telephone in the spring and summer of 2012.[3] Interviews addressed employer experiences with employees who are RC members, employer perspectives on USERRA and support programs provided by ESGR, and employer perspectives on the utilization policies and outreach efforts of the RCs. The willingness of employers to participate in these interviews was very low. Detailed information about the sample selection, participation rates, and recruitment for interviews and focus groups and the content of the interviews are provided in Appendix E, published separately online.

Interviewees are identified according to employer size and sector. The responses in the analysis regarding employer impact and employer awareness of USERRA and ESGR programs were generally grouped according to their similarity. The analysis also highlights congruence of thematic responses among similar interviewees (e.g., small firms versus large firms, or according to sector and industry) or lack thereof. Among interviewees who had experience with RC employees, the most-recurrent themes were timeliness of deployment notice, administrative processes related to RC absence and return to the workplace, and the multifaceted cost of increased or prolonged deployment of RC members.

Interviews with Reserve Component Chiefs

We interviewed each RC chief in 2011.[4] Interviews addressed the current state of and factors influencing the relationship between each chief's RC and the employer community, special programs that that RC has put in place to enhance relationships

---

[2]   Information from DMDC, 2009.

[3]   This data collection was approved by the Office of Management and Budget (OMB) (Control 0704-0484) in accordance with the Paperwork Reduction Act, Pub. L. 96-511, 1980.

[4]   We interviewed the chiefs (or their designees) of all seven RCs: ARNG, USAR, U.S. Navy Reserve, U.S. Marine Corps Reserve, ANG, AFR, and the U.S. Coast Guard Reserve. We also interviewed the chief of the National Guard Bureau, for a total of eight interviews. Most interviews took place between May and August 2011; one took place in November 2011.

with employers, and that RC's perspective on whether employers are upholding their USERRA obligations. These interviews were used to gather information on DoD policy and to shape the data analysis and interview data collection.

Limitations

This multimethod study draws on prior research and interview and survey data. The information on employer impact and on the experiences of RC members reported here reflects the perceptions of those who chose to respond to the survey or participate in the interviews. Although we and DMDC undertook extensive efforts to secure participation in the survey and interviews, participation rates for the surveys and interviews were low. Survey data have been weighted to correct for any sample selection bias based on observable characteristics, including state, employer size, employer industry, and employer sector. However, one must still be concerned that those who chose to respond to the survey are not representative of all employers or RC members. For example, employers that strongly support DoD or have had particularly good or bad experiences with RC employees may have been more likely to respond. We have no way to scientifically assess the nature of the potential bias from nonresponse. Given our experience screening participants, we can say that employers were not inclined to give up time to discuss an issue that was not especially salient to their organizations on a day-to-day basis.

Organization of This Report

The rest of this report is organized into five chapters. In Chapter Two, we present an overview of USERRA's provisions, origins, and legal history. In Chapter Three, we provide a framework for understanding and analyzing the impact that duty-related absences have on employers. Chapter Four presents our analysis of employer impact. Chapter Five describes employer perspectives of programs or efforts to support them in the event of duty-related absences. Chapter Six offers conclusions and recommendations.

A separately published set of technical appendixes (S. Gates et al., 2013) provides information for readers interested in the details of the analyses supporting this research. Appendix A provides a more detailed overview of USERRA. Appendix B presents our analysis of the DoD National Survey of Employers, and Appendix C presents an analysis of the January 2011 SOFS-R. Appendix D provides a focused analysis of findings organized by RC. Appendix E provides detailed information on the sample selection and recruitment for the employer interviews and focus groups that we conducted.

CHAPTER TWO

USERRA Overview

This chapter provides a brief overview and summary of USERRA as enacted and amended by Congress, as of January 2012. USERRA was signed into law on October 12, 1994, and since then has been revised multiple times (1996, 1998, 2000, 2004, 2008, 2011). Here, we highlight key provisions of the law as of January 2012.

USERRA combines aspects of antidiscrimination law with special reemployment rights for members of the armed forces and other selected beneficiaries. USERRA covers those who serve in or have served in the uniformed services, and it applies to all employers in the public and private sectors, including federal employers. For members of the RCs who leave an employer, USERRA provides valuable reemployment rights. To take advantage of USERRA's reemployment provisions, a service member must take care to comply with employee responsibilities, such as notification to the employer within a specified time frame. If the service member complies with these requirements, the service member is entitled to reemployment with his or her previous employer and may receive benefits as if he or she had never left the job for military service.

In addition to providing protection, USERRA establishes a system for pursuing legal claims that might arise from disputes between service members and their employers. These systems differ based on the type of employer and the choice of legal representation requested by the service member.

Appendix A examines the legislative and legal process history related to USERRA to provide better insight into the original intent of the law, as well as its implementation. We consider the effects of the legislation, as well as controversies that arose prior to and since its passage. The historical and legal contexts provided here can enhance DoD insight into the present-day standing of the law and its application and enable the department to ask the right kind of questions to appropriate USERRA stakeholders in the evaluation of USERRA's current functions.

USERRA Applies Broadly to Service Members and Other Selected Beneficiaries

USERRA's provisions apply to members of the Active and Reserve Components of the U.S. armed forces, the U.S. Public Health Service Commissioned Corps, and other categories of persons designated by the president in time of war or national emergency

(38 U.S.C. §4303[16]).[1] In some situations, coverage is even provided to veterans and those who apply to be in the uniformed services (38 U.S.C. §4311[a]). Furthermore, USERRA applies to all voluntary and involuntary uniformed service duty (38 U.S.C. §4303[13]). This includes but is not limited to weekend drills, annual training, active duty, and some local emergencies that trigger federal responses. USERRA covers a broad class of service; however, it has exceptions as well. The most important exception made is for discharge status: The law does not cover returning or former service members who receive unfavorable discharge from service (38 U.S.C. §4304[1][2]).[2]

For the remainder of this document, the term *service member* is used to refer to all beneficiaries of USERRA, regardless of their status in the armed forces or the Public Health Service Commissioned Corps.

USERRA Protects Service Members from Employment Discrimination

Generally speaking, USERRA is a law that protects service members from discrimination based on military service and provides further reemployment rights for service members returning to the civilian workforce. USERRA prohibits all employers, regardless of size, from discriminating against service members in hiring or routine matters of employment.

According to §4311(a), employers are prohibited from making discriminatory employment decisions because of one's military service (past, present, or future) or because one is applying to serve.[3] For example, an employer is in violation of USERRA if it refuses, because of an applicant's military service, to hire a veteran, an active member of the reserves, or someone in the process of enlisting. In this sense, the antidiscrimination protections are akin to those that exist in other statutes based on race, gender, and religion. Additionally, §4311(b) bars employers from retaliating against

---

[1] USERRA's benefits do not extend to service members who were dismissed for dishonorable conduct.

[2] Such discharges include service members who

(a) Separated from uniformed service with a dishonorable or bad conduct discharge;

(b) Separated from uniformed service under other than honorable conditions, as characterized by regulations of the uniformed service;

(c) A commissioned officer dismissed as permitted under 10 U.S.C. 1161(a) by sentence of a general court-martial; in commutation of a sentence of a general court-martial; or, in time of war, by order of the President; or,

(d) A commissioned officer dropped from the rolls under 10 U.S.C. 1161(b) due to absence without authority for at least three months; separation by reason of a sentence to confinement adjudged by a court-martial; or, a sentence to confinement in a Federal or State penitentiary or correctional institution. (38 U.S.C. §4304[1][2])

[3] Section 4311(a) covers a broad class of persons. It states,

a person who is a member of, *applies to be a member of,* performs, has performed, applies to perform, or has an obligation to perform service in a uniformed service shall not be denied initial employment, reemployment, retention in employment, promotion, or any benefit of employment by an employer on the basis of that membership, *application for membership,* performance of service, application for service, or obligation. (Emphasis added.)

employees who have brought USERRA claims or have participated in investigations involving a USERRA claim. This section covers people who have never participated in the uniformed services as well.[4] For example, an employer cannot fire or demote an employee who has made a USERRA complaint to DOL. In addition, an employer cannot fire a non–service member employee because the employee provided testimony in support of another's USERRA claim.

USERRA Entitles Employees to Reemployment After an Absence Due to Military Service

In addition to nondiscrimination protections, USERRA provides service members the right to return to their previously held jobs that they would have held with reasonable certainty had they not left for military service. Although the principle of reemployment is straightforward, the application of the law is more complicated. Issues concerning skill maintenance, promotions, health benefits, and paid leaves of absence are a few of the complications that may arise when determining the benefits to which a service member may be entitled.

The discussion in this section presents the responsibilities of the returning service member and exceptions to the general presumption of reemployment, then outlines the key aspects of reemployment benefits contained in the law.

Employees Seeking Reemployment Must Meet Five Requirements

An employee who leaves a job to serve in the military is entitled to reemployment under §4312. To qualify for reemployment rights, the service member must meet five legal requirements. First, an employee must provide his or her employer with written or oral notice that he or she will be leaving the job for military service–related reasons (38 U.S.C. §4312[a][1]). When giving that notice, the employee does not have to indicate whether he or she intends to return to employment.

Second, the cumulative length of absence due to military service cannot exceed five years (38 U.S.C. §4312[a][2]). The five-year clock applies to a single employer; a service member who changes employers is entitled to the five-year protection while working for each employer. Section 4312(c) provides for some significant exceptions to the five-year rule. For instance, mobilization due to war or national emergency does not count toward the five-year limit. Voluntary duty in support of certain operational missions or missions that have been designated by the secretaries of the military branches as "critical" are exempt from the five-year limit as well. Regular reserve duties, such as weekend drills and annual training, are also excluded. These significant exceptions lead one to ask, what purpose does the five-year clock serve to employers? The purpose of the five-year clock is to relieve employers of any legal obligations to service members who serve beyond five years of an initial obligation yet return for employment.

---

[4] Section 4311(b)(3) states, "The prohibition in this subsection shall apply with respect to a person regardless of whether that person has performed service in the uniformed service."

For example, an employee can enlist for active-duty service and serve an initial obligation up to five years and still receive USERRA protection. However, if this employee decides to reenlist for active duty, he or she will no longer be protected because he or she will have exceeded the five-year limit. This is consistent with USERRA's intent to protect noncareer service because, at the point of reenlistment, the employee's military service begins to resemble a career.

Third, upon completion of military duty, the employee must notify his or her employer, within a specified period dictated by the law, that he or she seeks reemployment (38 U.S.C. §4312[a][3]). The length of service determines the time frame for notifying the employer of the intent to be reemployed. The law prescribes the following time frames:

- If the employee serves fewer than 31 days, the employee must return to his or her employer by the next scheduled shift or calendar or after eight hours of rest and travel time. For example, if an employee returns from a two-week duty on a Saturday and his or her workweek normally resumes on Monday, then he or she should submit a reemployment application Monday morning (or whenever his or her shift normally resumes that day).
- If an employee serves for more than 31 days but fewer than 181 days, he or she must submit an application to the employer within 14 days after completion of service (38 U.S.C. §4313[a][1][A][B]).
- If an employee serves for more than 180 days, he or she must submit an application for reemployment within 90 days after completion of service (38 U.S.C. §4312[a][2][A][B]). This time provision is pertinent to reservists serving in OIF or OEF. Employees who seek reemployment after completing a four-year active-duty military contract will fall under this time provision as well.

These time frames and notification requirements are responsibilities of which employees must be mindful in order to secure reemployment rights. Failure to comply may result in a denial of their previous positions without recourse to the courts. DoD and ESGR have programs in place to notify employees of their rights and responsibilities under the law.

Fourth, the employee's absence must be due to service in the military. And fifth, the service member must not have received a disqualifying discharge from service.

Exceptions to the General Presumption of Reemployment Rights
Unlike other reemployment laws that exempt employers below a specified size threshold, USERRA applies to all employers regardless of their size. However, §4312(d)(1) provides exceptions that allow employers to deny reemployment to returning service members. As an example, an employer does not have to reemploy a service member if conditions in the business have changed so much that doing so would be unreasonable or impossible. Another exception involves "undue hardship," in that an employer does

not have to reemploy a service member if doing so would impose an undue hardship on the business (38 U.S.C. §4312[d][B]). *Undue hardship* is also a term of art.[5] The exception might apply in situations in which a service member incurred a disability while serving (after making any accommodations that are possible) or in which a service member is no longer qualified for the position that he or she left or would have held had he or she never left for reasons other than disabilities (after making attempts to make the service member qualified, per USERRA). In keeping with other antidiscrimination laws, USERRA uses the same definition of *undue hardship* as the Americans with Disabilities Act (Pub. L. 103-336, 1990).[6]

Another exception involves employers of temporary workers. Such employers do not have to reemploy service members who held positions that last for brief, nonrecurrent periods when there is no expectation of indefinite or significant continued employment. Courts have stated that whether a position is temporary depends on whether the service member had a reasonable expectation that his or her employment would continue for a significant or indefinite time.[7]

USERRA Protects Promotions and Job Advancement

Assuming that the service member notified his or her employer of the intention to return to work within the legally specified time frame and that the exceptions to presumptive reemployment discussed above do not apply, the service member will be eligible for reemployment with the employer with some of the rights and privileges that accrued as if he or she had never left. Determining the appropriate benefits and privileges is the subject of the following discussion.

Military service may require employees to be absent from their jobs for extended periods. To ensure that service members can retain their seniority and accrue some of the benefits of their employment tenure, USERRA includes protections for career advancement and promotion. Sections 4313(a)(1)(A)–4313(a)(2)(B) codify a principle known as the *escalator rule*, which uses the image of an escalator as a metaphor for career advancement.[8]

When a service member leaves for duty, he or she steps off the regular "escalator" of promotions and job advancement opportunities that occur during normal periods of

---

[5]   That is, a standard legal term referring to a provision that requires judicial interpretation when applied to a specific set of facts.

[6]   Title 38 U.S.C. §4302(15) and 42 U.S.C. §12111(10) are identical in their definitions of *undue hardship*.

[7]   See *Stevens v. Tennessee Valley Authority*, 687 F.2d 158, 163 (6th Cir. 1982) (in a case involving a USERRA predecessor, a service member who worked on a construction project that lasted almost a decade was denied reemployment because his job title was listed as temporary. The court held that a fixed term of employment is not necessarily temporary and, as long as a service member "has a reasonable expectation of reemployment on a regular basis in the future, [he or she] enjoys a right to reemployment").

[8]   The term *escalator rule* was coined by the U.S. Supreme Court in *Fishgold v. Sullivan Dry Dock & Repair Corporation*, 328 U.S. 275, 284–285 (1946).

employment. Missing out on these opportunities while on duty is a clear disadvantage to the service member and may be a disincentive to volunteer for military service. This disincentive would apply both to civilians considering joining the reserve and to current reservists who might be considering volunteering for a tour of duty.

To minimize the disruption that service would have on a civilian employee's career, the law extends reemployment protections to include the employee's right to those escalator advancements. At the completion of military service, the service member is entitled to reemployment at the point on the escalator he or she would have occupied had he or she kept his or her position continuously while serving. Thus, if a service member deploys for OEF for a year, he or she must be reemployed in whatever position he or she would have occupied, with a reasonable certainty, had he or she never left.

Upon the service member's reemployment, §4316 applies the escalator rule to benefits derived by seniority. For example, if a company policy grants an additional two days of vacation to employees who have worked for the firm for more than five years and an employee has been with the same company for five years, including one year away serving in Iraq, that employee would still be entitled to the additional days of vacation. Technically, the service member worked for only four years; however, the escalator rule ensures that the service member is considered to have worked for five years when it comes to benefits based on seniority.

A notable exception to this rule bears mentioning. The U.S. Supreme Court, in a case involving one of USERRA's predecessors, held that any benefit granted as a form of compensation for work that is actually performed is not protected because it is not derived from seniority status (see *Foster v. Dravo Corp.*, 420 U.S. 92, 99–100, 1975). For example, if an employee's terms of employment specify that he earns vacation based on hours worked, and the employee is absent for one year due to military service, he would not be entitled to the additional amount of vacation for the one-year hiatus. This is because the employee earns vacation only as compensation for hours worked and not based on total time employed.[9]

Determining the appropriate position on the escalator will depend on the length of time the service member has been away from the job. Time away from the job, however, may adversely affect the employee's skill set, either because the employee has lost some skills from disuse or because the work formerly performed has changed (because, for example, of newly adopted technology). Because of this, the service member's right to the escalated position is not absolute. It is important to note that the employer should make reasonable efforts to qualify a returning service member to fill his or her

---

[9]   Vacation days may be taken for the purpose of military service, at the employee's discretion. Section 4316(d) allows service members to use any paid vacation or leave they accrued prior to duty while serving in the military. This is done at the employee's election. Employers cannot make service members use vacation or leave to cover their absences. For example, an employer cannot require its service member employees to use vacation time while they do their two weeks' annual training with the reserves. However, a service member employee can use vacation time during his or her two-week annual training if he or she so chooses.

escalator position if time away from employment has resulted in the erosion or absence of necessary skills.

To determine whether a service member has a right to an escalated position, the employer must assess whether the employee has the requisite skills for the higher position. If so, then the service member should be employed in the higher position upon notification of desire for reemployment. If the employer determines that the service member is not qualified for the escalated position, then the employer must determine whether the employee can be trained in the new or missing skills with reasonable effort. If so, then the employer must expend the reasonable effort to train the returning employee.

If, however, the employer determines that the returning service member neither possesses the requisite skill for the escalated position nor could be trained with reasonable effort, then the employer's course of action will depend on the length of time the service member was away from the job. If the military duties lasted fewer than 91 days and the employer determines that the service member is no longer qualified for the escalated position and cannot become qualified after reasonable efforts have been made, then the employee must be reemployed in the same position he or she held when he or she left (38 U.S.C. §4313[a][1][B]). If the military service lasted for more than 91 days and the employee is neither qualified for the escalated position nor can become qualified after reasonable efforts have been made, then the employee can be reemployed in the position he or she left or one of like seniority, status, or pay (38 U.S.C. §4312[a][2][B]). Thus, the employer has some additional flexibility as to where in the company it can place a returning service member who has served for more than 91 days.

In sum, USERRA protects returning service members' interests in career advancement and seniority in terms of both benefits and position. But these protections are subject to scrutiny and a case-by-case determination.

USERRA Also Protects Health Plans and Pension Benefits

Service members are also provided protections for the benefits associated with their civilian employment, such as health plans and pension benefits. Service members have the option of keeping the health plans provided by their employers (38 U.S.C. §4317[a][1]). If someone serves less than 31 days, he or she cannot be required to pay more than the standard employee share of the coverage. However, if he or she serves for more than 31 days, the employer can require him or her to pay up to 102 percent of the full premium (including the share for which the employer typically pays) (38 U.S.C. §4317[a][2]). In addition, employers are required to offer access to employer-sponsored health plans to employees who are on military leave for only limited periods. The maximum period of coverage that employers are required to offer under USERRA is the lesser of either 24 months after the start of the service member's absence or the period of time up to the day after the service member fails to notify his or her employer of an intent to return to work (38 U.S.C. §4317[a][1][A][B]).

Regarding pension benefits, service member employees can maintain their employer-backed pension plans while away on duty. Their time spent away from their employers will be treated as employment time spent with their employers for accrual of pensions (38 U.S.C. §4318[a][2][B]). However, while a service member is on duty, the employer does not have an obligation to contribute to any employee- or employer-based contribution pension plans (such as a 401[k]) until the service member returns (38 U.S.C. §4318[b][1]).

Additional Reemployment Provisions Apply to Employees of the Federal Government

Reemployment with the federal government is covered under §4314 and is similar to reemployment with nonfederal employers. In certain situations, the Office of Personnel Management (OPM) plays a direct role in ensuring reemployment with the federal government. The OPM must employ service members with another federal executive agency in accordance with §4313 if it is impossible or unreasonable to reemploy them with the agencies for which they previously worked (38 U.S.C. §4318[b][1][A][B]). This rule also applies to service members who were employed by the legislative and judicial branches of the federal government (38 U.S.C. §4318[c]).

There are some federal agencies that are exempt from §4314 (38 U.S.C. §4315[a]). Some agencies, such as the Central Intelligence Agency (CIA) and the Federal Bureau of Investigation (FBI), are given some degree of discretion to determine whether reemployment under §4313 is practical (38 U.S.C. §4315[b]). Furthermore, reemployment decisions made by these agencies are not subject to judicial review (38 U.S.C. §4315[c][3]). Thus, any decision made by one of these agencies cannot be appealed in a court of law. This is more than likely due to the sensitive nature of the work performed by these agencies.

Key Employee Positions May Be Relieved of Some Duties to Prevent Employer Hardship

The foregoing discussion highlights the rights enjoyed by RC members when seeking a return to their civilian employment positions. However, it is important to note that employers have some limited capability to request the removal of an RC member from the Ready Reserve. Title 32 CFR §44 allows federal agencies to designate "key positions" that reservists shall not occupy. Any reservist holding a key position is to be transferred to the Standby Reserve (see 32 CFR §44.4[g]). Private employers may also seek to have key employees transferred to the Standby Reserve, effectively limiting the possibility that the RC member will be called for duty.

Title 32 CFR §44 creates guidelines for determining whether an employee is in a key position. The guidelines place a strong emphasis on designating positions that are vital to emergencies, public health, and safety (32 CFR §44). However, these guidelines are not mandates; they do not have to be followed, but employers are encouraged to do so (32 CFR §44). In the end, it is the secretary of each branch of service who decides whether a reservist is transferred to the Standby Reserve (see 32 CFR §44.5[c][5]).

We discussed this provision with staff from the Office of the Assistant Secretary of Defense for Reserve Affairs (OASD[RA]). Since 2008, only 275 requests to have key personnel exempted from a mobilization were received. Thirty-five percent of these requests came from federal agencies; 65 percent were from nonfederal sources. Of all requests made, 95 percent were granted, resulting in either a negation of the RC member's call to duty or a delay in the mobilization of up to six months.

Enforcement of USERRA Ensures Service Members' Coverage

There are two federal programs that help resolve USERRA-related disputes between service members and employers. First, any service member who believes that his or her rights under USERRA have been violated can make a complaint to DOL's VETS. VETS aims primarily to investigate employment disputes before legal action is taken, though it also mediates some.

Second, if an employee is a member of the Guard or Reserves, the employee has the option of making a complaint with ESGR. ESGR provides informal mediation services through its ombudsmen. ESGR strives to resolve all termination complaints and non-reinstatement issues between employers and service members within seven business days (DOL, 2008, p. 3). This is just an option, not a required step that reservists must take to resolve USERRA issues. They are free to use VETS or bring lawsuits through private attorneys.

At all times, a service member can bring a claim on his or her own through a private attorney (or self-representation). Service members do not necessarily have to seek assistance through VETS or ESGR; however, service members are required to go through VETS if they seek assistance from either the U.S. Attorney General (AG) or U.S. Office of Special Counsel (OSC) (38 U.S.C. §4323[a][1]; 38 U.S.C. §4324[a][2][A]). There are also procedural guidelines that must be followed depending on whether the employer being sued is a private company, federal agency, or state government. These procedures are summarized in this section.

Private Companies

If a service member is employed by a private company, he or she has two options: seek private counsel or request assistance from the AG (38 U.S.C. §4323[a][1], [3]). For a service member to request assistance from the AG, he or she must first go to VETS. If VETS cannot resolve the issue, the service member can have DOL refer the claim to the AG (38 U.S.C. §4323[a][1]). If the AG reasonably believes that the service member has a claim, the AG may appear on behalf of the service member and bring the claim in a federal district court (38 U.S.C. §4323[a][1]; 38 U.S.C. §4323[b][1]). Alternatively, if a service member uses private counsel, the attorney can bring the claim in a federal district court (38 U.S.C. §4323[b][3]). See Figure 2.1 for a diagram of the process involved in making a claim against a private company.

Federal Agencies

If a service member is a federal employee, he or she must have the claim adjudicated by the Merit Systems Protection Board (MSPB) (38 U.S.C. §4324[c][1]). A service member can also request assistance from OSC, but he or she must go through VETS first (38 U.S.C. §4324[a][1]). If VETS cannot resolve the issue, the service member can have DOL refer the claim to OSC (38 U.S.C. §4324[a][1]). If OSC reasonably believes that the service member has a claim under USERRA, OSC may appear on behalf of the service member in front of MSPB (38 U.S.C. §4324[a][2][A]). MSPB's decisions can be appealed to the federal circuit court (38 U.S.C. §4324[d][1]). See Figure 2.2 for a diagram of the process for making a claim against a federal agency.

State Governments

A service member employed by a state government cannot sue his or her employers in a federal court unless the state waives its right to sovereign immunity (38 U.S.C. §4323[b][2]). However, if the AG brings a suit on behalf of a service member against a state, the federal courts will have jurisdiction (38 U.S.C. §4323[b][1]). For example, if the AG is willing to bring a claim against a state university that violated one of its employee's USERRA rights, the lawsuit can come before a federal district court and the claim will be brought on behalf of the United States (38 U.S.C. §4323[a][1]). In contrast, if the employee sues the state university through private counsel, the lawsuit must go before a state court. See Figure 2.3 for a diagram of the process for making a claim against a state government.

Summary of Enforcement Options

In sum, when it comes to enforcing USERRA's protections, service members are not without options. In addition, if a service member chooses to bring his or her own claim and prevails, the court may award attorney fees and litigation expenses (38 U.S.C. §4323[h][2]). Furthermore, USERRA prohibits claimants from being charged fees and court costs (38 U.S.C. §4323[h][1]). These enforcement guidelines and procedures ensure that the U.S. civil justice system is readily accessible to service members and that there will be few impediments in the way of obtaining justice.

States may also have state-specific legislation that provides protection for service members. These state-level antidiscrimination and reemployment laws can supplement and exceed the level of protection contained in USERRA. State laws, however, may not provide less protection than that afforded under USERRA.

Figure 2.1
Process Flow for Making a Claim Against a Private Company

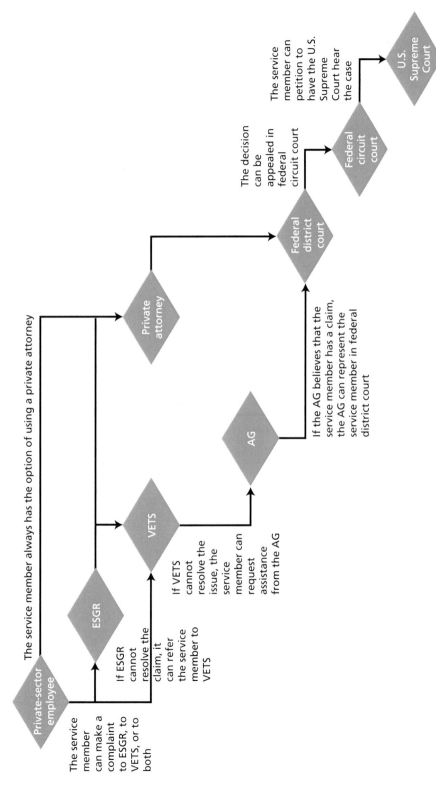

The service member always has the option of using a private attorney

Private-sector employee

The service member can make a complaint to ESGR, to VETS, or to both

ESGR

If ESGR cannot resolve the claim, it can refer the service member to VETS

VETS

If VETS cannot resolve the issue, the service member can request assistance from the AG

Private attorney

AG

If the AG believes that the service member has a claim, the AG can represent the service member in federal district court

Federal district court

The decision can be appealed in federal circuit court

Federal circuit court

The service member can petition to have the U.S. Supreme Court hear the case

U.S. Supreme Court

Figure 2.2
Process Flow for Making a Claim Against a Federal Agency

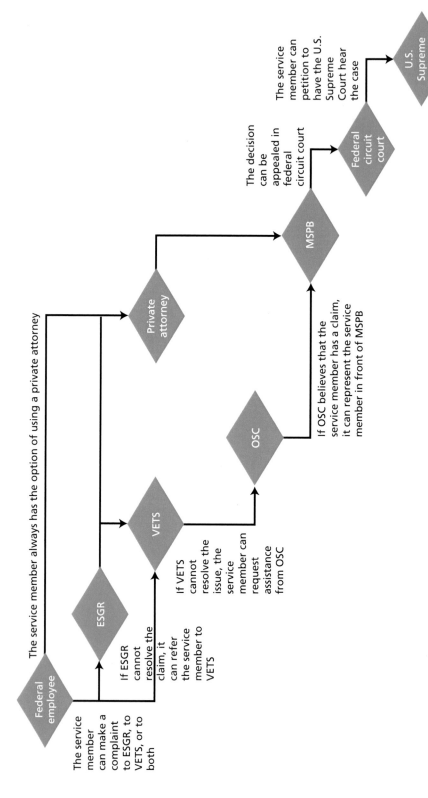

The service member always has the option of using a private attorney

Federal employee

The service member can make a complaint to ESGR, to VETS, or to both

ESGR

If ESGR cannot resolve the claim, it can refer the service member to VETS

VETS

If VETS cannot resolve the issue, the service member can request assistance from OSC

OSC

If OSC believes that the service member has a claim, it can represent the service member in front of MSPB

Private attorney

MSPB

The decision can be appealed in federal circuit court

Federal circuit court

The service member can petition to have the U.S. Supreme Court hear the case

U.S. Supreme Court

Figure 2.3
Process Flow for Making a Claim Against a State Government

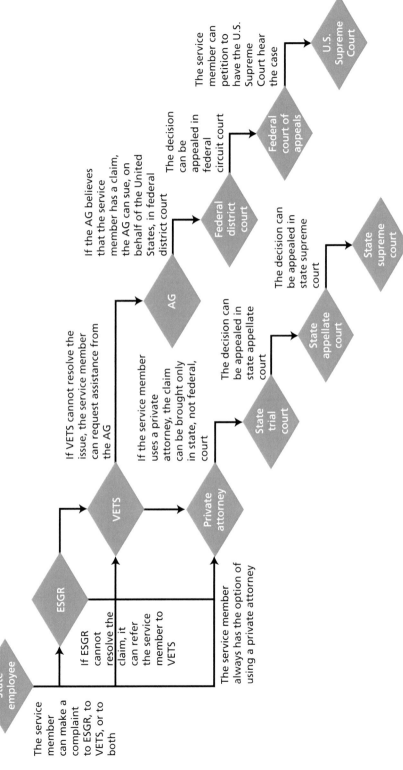

Court Interpretation of USERRA

The implementation of USERRA has presented the courts with multiple opportunities to examine the USERRA law for clarity, consistency, and adaptability to changing employment practices. As part of this study, we reviewed appellate court decisions related to USERRA. That review is described in Appendix A. The legal history surrounding the implementation of USERRA reveals several areas in which the law may be presenting difficulties to service members and employers alike. Appellate courts have seen an increase in USERRA cases since 9/11. From a review of these cases, we summarize, in Appendix A, issues that are arising with greater frequency and that present topics of current salience—a judgment based on our view of the legislative and case history review.

Current issues arising in court include questions about whether a service member can accept, and whether an employer can offer, waivers of USERRA rights in exchange for an alternative benefit (namely, money). Because USERRA allows employers to provide (and states to require) benefits to service members that exceed the level provided under the federal law itself, the courts have wrestled with the question of whether negotiated lump-sum payments and contracts specifying a preference for arbitration (rather than litigation) are benefits above the federal threshold. These severance payments raise questions about equitable bargaining power between employer and employee, as well as issues about the proper valuation of USERRA rights in exchange for money. Although it is not clear how many such cases are likely to arise in the future, the courts are well equipped to resolve these types of issues.

More-frequent controversies have arisen presenting questions about arbitration of USERRA claims. Employers are increasingly asking employees to agree, in writing, to arbitration procedures in the event of a legal dispute with their employers. USERRA claims are just one of many types of claims that are covered under such an agreement. Here, the circuit courts have reached different conclusions, a situation that may indicate future U.S. Supreme Court intervention or congressional direction. Some circuits have found a procedural right to a day in court, while others have determined that no such substantive right exists and endorsed arbitration as an acceptable substitute. The result of these different decisions complicates an already multimethod approach to bringing USERRA cases. The situation presents multiple venues for service members to pursue (and employers to defend) their claims. RAND has not been asked to study and recommend a preferred procedure for hearing USERRA complaints; however, such a review holds the potential to identify efficient and streamlined processes for resolving these unique employment controversies.

An area of controversy that was resolved recently (while we conducted this study) was a split in interpretation arising when courts address whether service members have a right to claim employment discrimination under USERRA on the grounds of hostile work environment. Here, the root cause of the controversy was the lack of statutory language specifically protecting service members from hostile environments. However, legislative revision in late 2011 clarified congressional intent by harmonizing the USERRA language with that of similar antidiscrimination laws (Pub. L. 112-56). Such revision was recommended by DOL in a 2010 report to Congress (DOL, 2011).

Summary

USERRA was written to encourage service in the U.S. armed forces by removing some barriers to military service that are related to civilian employment. In its current form, there are two main aspects of the law: (1) general antidiscrimination protection based on veteran status and (2) reemployment rights after periods of military duty. Both categories of protection offer valuable benefits to veterans, as well as to civilians considering joining the military. In passing USERRA, Congress recognized that these rights could impose costs on employers. Congress considered whether employers should be exempt from USERRA based on size and did not include such an exemption in the final legislation.

The history of service member antidiscrimination law is long and somewhat complicated, but the current USERRA law maintains the basic tenets of the earliest laws. Namely, a service member has the right to seek reemployment with his or her civilian employer within a specified time frame after returning from duty; the service member must meet certain responsibilities concerning employer notification (both of leave for duty and of intent to be reemployed), and service member rights include protections for the type of position to which the service member is entitled upon his or her return.

CHAPTER THREE

Perspectives on the Impact That Military Duty–Related Absences Can Have on Employers

As described in the previous chapter, USERRA conveys specific employment and reemployment rights and benefits on veterans and members of the RCs. When an RC member is activated, the employer loses an employee and is thus affected. Those effects could be harmful or beneficial to the employer. Although it is common to describe those impacts as "effects of USERRA," many employers voluntarily provide benefits to activated RC members that exceed those required by the law. As such, we describe these impacts as the effects of duty-related absences in the context of USERRA. In this chapter, we provide a conceptual framework for understanding the impact that military duty–related absences can have on employers. We describe the factors that likely influence the magnitude and direction of that impact, drawing on prior research on USERRA and other employment laws—in particular, the FMLA as the closest analogue to USERRA.

Conceptual Framework Linking Military Duty–Related Absences to Employer Impact

A 2005 report from the Congressional Budget Office (CBO) highlighted the potential employer burden resulting from the operational reserve (Golding, 2005). Based on data available at that time and supplemented by qualitative information obtained through interviews with 22 employers, the report describes the potential burden of military absences on employers and the USERRA-related costs in the event of a reservist call-up. CBO categorized the sources of the cost burden, which include the short- and long-term costs of filling the vacancy, disruptions in normal business activity (which may include work slow-downs or, in the extreme case, a business closure), and the costs of continuing benefits (especially retirement benefits), as well as the costs associated with dealing with a USERRA complaint. CBO also suggested that the cost of an activation for a business is likely to vary depending on characteristics of the business, the position that is filled by the RC member, and how long the RC member is away from work.

The themes highlighted by CBO with respect to USERRA's potential impact on employers are echoed in the broader literature on workplace absenteeism, which is said to have a high cost for employers. The costs of absenteeism include employee replacement costs, decreased productivity, decreased morale, and potential performance and safety issues when replacing a key employee (Gaudine and Saks, 2001; Navarro and Bass, 2006). The cost of unscheduled absenteeism in 2005 was $660 per employee, a $50 increase from the 2004 estimated cost ("Absence Survey Finds Costs Are Up," 2006). The direct and indirect costs of absenteeism could reach 15 percent of payroll, according to industry surveys (Navarro and Bass, 2006). Research also suggests that small firms (or isolated locations of larger firms) are expected to have higher costs of absenteeism because they are unable to easily replace absent workers with current staff (Pauly et al., 2002). The cost of an absent worker can be substantially higher than that worker's wage, given productivity losses of the affected team or division of the firm (Pauly et al., 2002).

Although we focus the remainder of this report on the impact that actual duty-related absences have on employers, it is important to acknowledge potential costs of USERRA that do not follow directly from an absence but rather stem from the mere existence of the law. The first is the expected cost of potential litigation under USERRA (Golding, 2005), and the second is the recordkeeping and documentation burden that follows from the law and the potential for litigation. We uncovered no studies documenting or estimating the expected cost of litigation or the recordkeeping burden of USERRA. However, available information on complaints submitted to DOL and the disposition of those complaints suggests that the expected costs are not large.

DOL reviewed 1,576 new USERRA cases in FY 2011. OSC reviewed an additional 28 new cases as part of the USERRA demonstration project (DOL, 2011). The number of new cases reviewed by DOL and OSC was slightly higher (about 10 percent) than in prior years. DOL closed 1,675 cases in FY 2011. Sixty-three percent of those cases (1,055) were withdrawn or closed due to a lack of merit, eligibility, or interest. Employers granted the request in the claim or settled the claim in 24 percent (402) of the cases.[1] The remaining 218 claims were not resolved by DOL, and most of these were referred to the regional solicitor's office. In addition, the ESGR ombudsman handled 2,884 cases, resolving 80 percent (2,302) of those. The RC member in the unresolved cases had the option to file a case with DOL or to pursue a private claim. DOL estimates that 299 of the cases handled by ESGR were also counted as new cases by DOL.

Although firms may choose to keep records related to USERRA to assist with responding to potential claims, USERRA does not impose any recordkeeping or reporting requirements on employers. A Darby Associates team estimated that, in 2004, complying with the FMLA (which does include specific recordkeeping require-

---

[1] Golding (2005) reported that, for cases resolved in 2004, the average award by the employer to the RC member in granted or settled cases was about $2,500. However, the amount of the claim payment does not account for the legal and opportunity costs that could be incurred in a larger number of cases beyond those in which a payment was made.

ments) cost firms approximately $32 billion, including both direct and indirect costs of compliance. The authors concluded that much of this cost comes from employees "gaming" the system—taking advantage of the FMLA and using it for unqualified leave (Darby and Fuhr, 2007).

Our framework for understanding the impact of duty-related absences in the context of USERRA builds on two key insights in the prior literature. First, the net impact on employers stems from both direct and indirect effects of dealing with the absence in a way that complies with USERRA. Direct impacts are those associated with recruiting, screening, hiring, and training replacement workers, as well as retraining the reservist upon return from military duty (as needed), the relative cost of the replacement worker (which could be higher or lower than those for the absent reservist), and the cost of benefits that the employer provides to the employee during the absence (either voluntarily or as required by law). In addition, indirect costs, such as lost business, productivity, or opportunity for growth, may be incurred. Second, the impact will vary across employers and will also be influenced by which employee is being deployed and for how long.

Figure 3.1 describes our conceptual framework. By establishing employment and reemployment rights, USERRA has the potential to impose direct costs on employers. Those costs are influenced by characteristics of the employer, the absence, the overall economy, and the position of the employee who is activated. Activation poli-

Figure 3.1
Conceptual Framework Linking a Duty-Related Absence to Employer Impact

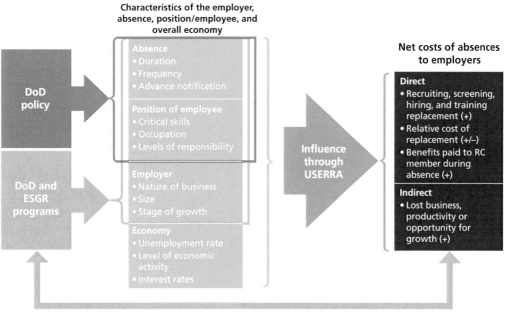

cies and practices influence who is called to duty, how often, and for how long. This can influence costs indirectly through and have an effect on the nature of absences, the characteristics of employees who are activated, and potentially the characteristics of employers. DoD and ESGR programs to support RC members and employers can have a direct influence on some of those costs. We discuss each category in this chapter.

Characteristics of the Employee and the Position

Golding (2005, pp. 17–18) notes that the activation of a highly skilled or specialized worker is likely to have a larger negative impact on the employer than activation of a less skilled or specialized worker. These points are drawn from basic insights about labor markets. This larger potential impact is due to several factors. An employer may have more difficulty replacing a highly skilled or specialized worker. It may need to undertake a costly or time-consuming search to identify a worker with the skills, certifications, or clearances required to do the job of the activated reservist. In some cases, the employer may simply not be able to find a temporary replacement before the activation occurs. Some employers may decide to hire a replacement worker who lacks the needed skills and provide the necessary training. Those training costs would be borne by the employer. An employer may be more likely to suffer lost productivity or loss of business when a highly skilled worker is activated, either because it is unable to find a replacement or because the replacement does not quite have the skill levels or firm-specific human capital of the activated reservist. To the extent that highly skilled or specialized workers are paid higher wages, the employers would incur higher retirement benefit costs during an activation.[2] Finally, some highly skilled workers may need to be retrained when they return from their military duty.

Characteristics of the Employer

In 2007, Doyle and Singer (2008) surveyed 478 employers of RC members about the cost that RC absences imposed on their businesses in 2005 and 2006. Employers included in the study reported employing RC members and were aware that at least one had been activated during 2005 or 2006. Most employers surveyed reported little or no activation-associated costs to the workplace. Within each category of employer (small business, large business, nonprofit, and state and local government), there was variation in reported costs. Some employers reported a financial benefit from the activation, while others reported costs in excess of $3,000 (Doyle and Singer, 2008, Figure 3, p. 139). State and local governments and small businesses were the most likely to report substantial costs, with roughly 20 percent of the organizations in each of these groups citing costs more than $10,000 (Doyle and Singer, 2008, Figure 4, p. 139): "All of the severely affected government agencies were first responders (FRs),

---

[2] Although employers do not pay employee wages during military leave, they are required to maintain retirement benefit accruals.

such as police departments or emergency medical teams. Their high costs stemmed from training and overtime wages for remaining personnel" (p. 140). Some state and local government agencies face restrictions on "backfilling" a vacant position when a person is on leave. Small businesses cited training expenses for replacement workers as the source of high cost.

Doyle and Singer (2008) also asked respondents about lost business. Five percent of large employers and 21 percent of small employers reported that they suffered business losses due to reservist activation, with reported losses ranging from $25 million to $1 million (p. 140).

In 2009, Hope, Christman, and Mackin (2009) used data from Dun and Bradstreet's business-sales database (the Data Universal Numbering System, or DUNS) and the DoD reservist-employer database to analyze the effect that reserve activation can have on small-business performance. The study revealed that the activation of RC employees is related to a negative effect on business sales overall and a larger negative effect on small businesses than on large businesses. The authors also found that the length of activation has a small but significant negative effect on a firm's revenues. Though numerous questions remain, including the quality of the self-reported data on small businesses in DUNS, this study represents one of the first attempts to evaluate the economic impact that reserve activation can have on a firm from the employer's perspective.

The costs of employee absences and their relationship to employer characteristics have been considered in the context of the FMLA. A 2002 report estimates the cost of expanding the FMLA to small businesses; firms with fewer than 50 employees are currently exempt from compliance with the FMLA, although some states' family and medical leave laws have lower thresholds. The author of that report concluded that expanding the FMLA to small businesses could impose costs of $30,000–50,000 through loss of sales, overtime and temporary worker payments, and administration of the act. He pointed out that the loss of a "key" employee likely affects small businesses more than large businesses because it is unlikely that a small business has someone else on staff with a skill set similar to that of the lost employee (Phillips, 2002).

A majority of respondents to a 2006–2007 survey about the FMLA reported increasing other employees' workloads when leave is taken (Leonard, 2007). Small businesses may be less able than large businesses to accommodate an absence in this manner.

Questions have also been raised as to whether federal agencies face particular challenges in dealing with USERRA obligations. President Barack Obama released a memorandum on July 19, 2012, directing the leaders of federal government agencies and executive offices to "take steps to ensure robust compliance with USERRA's employment and reemployment protections across the Federal government through outreach, education, and oversight" (Obama, 2012). The memorandum directed the establishment of the USERRA Employment Protection Working Group to coordinate

and review federal agency efforts related to USERRA, promote outreach and best practices, and improve the data on USERRA compliance. The memorandum was released
subsequent to media reports about large numbers of USERRA-related complaints filed
against federal agencies.

Although our interviews with RC chiefs were not focused on federal employers,
when we asked a general question about how the employer relationship varies across
types of employers, two of the eight interviewees mentioned the federal government as
a source of major challenges. One respondent said, "The federal government is in the
bottom tier of employers that understand that the folks they have are also members of
the military." Another respondent who said that the "biggest challenge we have is the
[federal] government" went on to suggest that the problems seem to stem from a failure on the part of central offices to communicate USERRA-related policies to lower
levels of the organization. However, another interviewee said that RC members who
work for "military-friendly organizations, such as the VA [the U.S. Department of
Veterans Affairs], police, and fire departments, usually don't have any problems with
reemployment."

The studies mentioned here indicate that most employers do not experience significant cost or productivity implications when employees are absent but that some
employers experience significant impacts. Although specific characteristics of employers have been associated with a greater likelihood and higher level of cost impact, the
fact that there is substantial variation in cost impact even within categories of employers suggests that the cost implications are often idiosyncratic in nature. In general, state
and local governments and small employers seem to be more vulnerable to employee
absences than are larger private employers. Concerns have been raised about federal
employers and their ability to live up to USERRA obligations.

Characteristics of the Absence: Advance Notification Provided, Duration, and Frequency

As discussed by Golding (2005), some of the costs associated with a duty-related
absence are one-time costs that are incurred regardless of how long the employee is
away. For example, if an employer chooses to hire a replacement worker, the hiring and
screening costs are borne one time regardless of whether the reservist is away for three
months or one year. Other costs are incurred over time while the reservist is away, such
as employer contributions to the retirement plan or the cost associated with overtime
wages for employees who are asked to assume the workload of the absent RC employee.

The relationship between cost and advance notification has not been explicitly
examined in the context of USERRA, but there is reason to expect that a lack of
advance notification would impose greater costs on some employers, especially in circumstances in which it is difficult to find replacement workers. Such a relationship
has been suggested in the literature on the cost of the FMLA. The most costly type of
leave under the FMLA is intermittent or short-term leave, according to a report pub-

lished in 2005 by the Employment Policy Foundation (EPF). According to that report, about half of FMLA leave-takers do not give advance notice of their leave, and this lack of advance notice may increase the impact that such leave can have on employers (Ganapati, 2005).[3]

Characteristics of the Economy

Although this issue has not been the subject of empirical studies, we would expect an employer to experience a different impact depending on whether an absence occurs during a recession or during a period of solid economic growth. During times of high unemployment, it may be relatively easy to find a temporary replacement or to shift work to other employees who are being underutilized. Because the employer does not need to pay the activated reservist during the duty period, the activation could even result in cost savings for the employer. In a tight labor market or one in which business is booming, employers may face high costs of replacing absent workers and may suffer economic loss due to lost productivity. The data used in our study were collected during a major economic recession. Although we are not able to test this hypothesis in our study, the reader should keep this potential relationship in mind.

U.S. Department of Defense Policy

How DoD and the RCs choose to use reservists can influence who is activated, how frequently, and for how long. Although there is some DoD-wide policy that constrains RC utilization, the RC chiefs still have substantial latitude over key utilization decisions. The specific activation policies and practices have varied across components and over time. In this section, we describe DoD-wide policy constraints and key utilization decisions that are made at the component level that might influence the likelihood, duration, and frequency of activation of RC members.

DoD policy limits involuntary activations to a ratio of one year activated to five years deactivated (R. Gates, 2007).[4] The amount of time that an individual is deployed within that activation limit varies by component. In the context of this policy, the Army has adopted a contiguous training model that combines scheduled annual training and weekend drill periods (up to 45 days) for the purposes of predeployment training to be conducted immediately prior to a deployment-related activation. From an employer's perspective, this model may result in fewer absences, but absences that exceed one year in duration.

Employers could interpret the 2007 policy as a guarantee that RC members would be activated for no more than one year in a five-year period. However, the policy

---

[3] This article, along with another in *HR Magazine* (Bates, 2005), summarizes the findings of a report published in 2005 by EPF; the original report could not be located because EPF has closed.

[4] Prior to 2007, the length of mobilizations was not limited by DoD policy, and the RCs, especially the ANG and USAR, did mobilize reservists for longer periods of time.

applies only to involuntary activations related to deployment (mobilizations). As discussed in Chapter Two, USERRA covers duty-related absences due to both voluntary and involuntary activations up to a five-year limit. As a result, RC policies and practices related to the use of voluntary versus involuntary activations will influence the pattern of military duty for RC members. These patterns can, in turn, affect employers, particularly those that employ reservists who volunteer regularly.

DoD policy can also play an important role in determining whether extended service by RC members is exempt from the five-year cumulative length of service limit. Two exemption categories are particularly relevant here. The first is the exemption under 38 U.S.C. §4312(c)(4)(C) for those who are called to active duty after volunteering to support an operational mission. The second exemption, under 38 U.S.C. §4312(c)(4)(D) applies to volunteers who are ordered to active duty in support of a critical mission or requirement. The secretaries of the military departments have the authority to designate a military operation as a critical mission or requirement. To the extent that RCs rely on volunteers and the secretaries exempt that service from the five-year limit, policy and practice may be concentrating the costs of duty-related absences on particular employers (specifically, those that employ individuals who volunteer for duty).

Some RC activation practices can influence who gets activated and for how long. From an employer perspective, variation in the activation practices across RCs can imply that employees in certain RCs or certain career fields are more likely to experience an activation. Table 3.1 shows that the Navy Reserve has the lowest percentage of recent activations, at 27 percent, while the ARNG has the highest, at 36 percent. At the time of this survey, 5 percent of all respondents were deployed, with the percentage by RC ranging from 3 percent in the Marine Corps Reserve to 7 percent in the USAR.

Table 3.1
Percentage of Reserve Component Members Activated in the Past 24 Months, by Component, January 2011

Status	ARNG	USAR	Navy Reserve	Marine Corps Reserve	ANG	AFR	All RC Members
Not activated	64	66	73	66	67	70	66
Activated	36	34	27	34	33	30	34

SOURCE: Authors' analysis of 2011 SOFS-R data.

NOTE: Margin of error ranges from ±1.0 percent to ±2.3 percent. Weighted Ns: ARNG = 343,829; USAR = 195,823; Navy Reserve = 61,829; Marine Corps Reserve = 37,718; ANG = 104,900; AFR = 66,889; all = 810,989. Unweighted Ns: ARNG = 4,462; USAR = 4,100; Navy Reserve = 3,548; Marine Corps Reserve = 2,586; ANG = 3,231; AFR = 3,724; all = 21,851. The last column reflects the overall survey population. Some respondents may not have reported their service affiliations. These are not intended to be averages of the prior columns, and the numbers should not average or sum.

Another aspect of RC practice that can affect rates of activation is the extent to which a military service relies on the RCs for certain skills and functions. Hansen et al. (2011) found that the USAR is the "most *unbalanced* of the components—that is, the extent to which service members in its high-utilization career fields are currently mobilized is disproportionately high relative to the component average" (p. xv). High-utilization career fields were identified as of December 2008 on the basis of a combination of measures that reflect the percentage of service members currently activated or ever activated and the amount of time they have been activated. For the USAR, the high-utilization career fields were civil affairs, transportation, chaplain, psychological operations, explosives and ammunition, infantry, law enforcement, and finance. For the ARNG, high-utilization career fields were infantry, air defense, field artillery, law enforcement, and armor (Hansen et al., 2011, Tables 5.1 and 5.2, p. 33). It is worth noting that law enforcement and infantry are high-utilization career fields in both the USAR and ARNG.

Whether an RC member is in a high-utilization career field can have significant implications for the frequency and duration of activations. As of December 2008, of RC members in the civil affairs career field in the USAR, 27 percent were activated and 78 percent had been activated at some point. RC members in this career field had spent, on average, 22 percent of their time activated since September 2001. In contrast, among those in the metalworking career field, 3 percent were activated, 30 percent had been activated since September 2001, and current RC members had spent an average of about 6 percent of their time activated since September 2001 (Hansen et al., 2011, Table B.2, pp. 69–71).

DoD policy also seeks to mitigate the difficulties experienced by employers that face employee absences arising from RC activations. DoD Instruction (DoDI) 1205.12, written in 1996 and last revised in 1997, contains an official policy concerning adjustments to reserve duties when employers' business interests could be adversely affected (Under Secretary of Defense for Personnel and Readiness, 1997). DoDI 1205.12 instructs the RCs to identify a representative who

> shall consider, and accommodate when it does not conflict with military requirements, a request from a civilian employer of a National Guard or Reserve member to adjust a Service member's absence from civilian employment due to service when such service has an adverse impact on the employer. (p. 8)

When such employer requests are received, the representative is directed to "make arrangements other than adjusting the period of absence to accommodate such a request when it serves the best interest of the military and it is reasonable to do so" (p. 8). Although the policy has been in place since the 1997 revision to the instruction, such details as how often such requests are made, granted, or denied remain unclear because there is no DoD recordkeeping program for the policy. Similarly, the standards for granting or denying requests for adjustments are not presently standardized across DoD. At the time this report was written, DoD was reviewing the instruction, and revisions were expected to be released in 2013.

The policy, data, and prior studies reviewed in this section suggest that the likelihood and duration of military absences will depend on the employee's RC, the employee's career field, and, importantly, the employee's propensity to volunteer. DoD-wide policy places some limitations on activations but leaves substantial room for RC practices and the decisions of RC members to influence these outcomes. Our analysis explores whether policy changes related to utilization should be considered; our findings and recommendations are summarized in Chapter Six.

DoD and ESGR Programs

DoD describes employers as the "third leg of the stool" (along with RC members and their family members) that supports individuals' ability to serve in the RCs. For this reason, DoD engages in extensive outreach efforts designed to educate, recognize, and support employers.

Since 2004, RC members have been required to provide DoD with information on their civilian employers and job skills and to notify DoD of any job changes (Chu, 2003; DoD, 2004). The data provided by RC members make up the CEI database described earlier. A primary motivation for this requirement was to support DoD's efforts to inform civilian employers about USERRA rights and responsibilities.

In establishing this requirement, DoD delegated responsibility for implementation to each military department, and separate CEI programs were established. The U.S. Government Accountability Office (GAO) conducted a review of the CEI programs in 2007 and found some limitations on the quality of employer information collected (GAO, 2007). Subsequently, DoD revised some of the processes for collecting employer information. It is worth noting that, although the CEI database can be used to promote general employer outreach, DoD does not allow the CEI program data to be used to contact employers in any way "that would identify the RC service member employed by that employer without the prior consent of such RC service member" (Under Secretary of Defense for Personnel and Readiness, 2011b, p. 57).

DoD sponsors multiple organizations and programs that support service members and their employers in their transitions to and from civilian employment and military duty and enhance awareness of USERRA and its provisions. ESGR is one such organization with this primary purpose. ESGR's vision is to create "a culture in which all American employers support and value the employment and military service of members of the National Guard and Reserve." ESGR's mission is to facilitate and promote

> a cooperative culture of employer support for National Guard and Reserve service by developing and advocating mutually beneficial initiatives; recognizing outstanding employer support; increasing awareness of applicable laws and policies;

resolving potential conflicts between employers and their service members; and acting as the employer's principal advocate within DoD. (ESGR, undated [d])

ESGR uses a nationwide network of volunteers to conduct informational briefings on USERRA, recognize exemplary employers whose policies and actions encourage military participation, and provide mediation services when issues arise. A summary of programs and initiatives by ESGR is provided in Table 3.2.

The chief of the USAR originally established the Employer Partnership of the Armed Services program (EP) in 2008 to promote collaboration between the RCs and employers with mutual interests, such as employer and DoD readiness for mobilizations and employment of RC members. Resources developed through the program were extended to the other RCs by 2010. According to Pint et al. (2012), "efforts through September 2010 have focused primarily on reducing unemployment among reserve component (RC) service members, improving linkages between military and civilian skills and occupations, and establishing closer working relationships with employers" (p. xi). Employer outreach is achieved through program support managers (PSMs) who are also responsible for outreach to RC members. The PSMs work to identify employer partners. Activities focus primarily on the hiring of RC members and helping employer partners deal with military training and employment (EP, undated).

In addition to ESGR and EP, there are service-specific and government-wide programs designed to foster linkages between civilian employers and DoD. Most of these are designed to promote employment opportunities for veterans, RC members, or military family members or to provide military members with certifications or academic credit for training or experience received while in the military. Examples include the Army's Partnership for Youth Success (PaYS) and Credentialing Opportunities On-Line (COOL) and DOL's VETS (Pint et al., 2012, pp. 5–7).

Golding (2005) suggested several programs that the U.S. government could institute to further support employers, including compensation through tax credits or direct payments, subsidized loans to employers, call-up insurance for businesses, or exemption from call-up for some reservists. The U.S. government does not currently offer financial assistance to employers of RC members in the event of a military absence, although governments in other countries do provide such assistance. Doyle and Singer (2008) argue that broad-based financial incentives are unwarranted because most employers do not experience cost implications. Instead, they suggest that, if a financial incentive program were implemented, it should take the form of a targeted grant program. As discussed in Busby (2010), broad programs have the advantage of being easy to administer and require no effort on the part of employers. However, they may be costly to the government and unfair to individual employers. Targeted programs allow the government to compensate those employers that do suffer losses, but they impose a significant burden on employers and the government and open the government to fraudulent claims. Our analysis explores the effectiveness of existing

Table 3.2
Summary of Employer Support of the Guard and Reserve Programs

Title	Description	Notes
SOS document	Document signed by an employer as visible evidence of its support for the Guard and Reserve. This serves as an acknowledgment of its responsibilities under USERRA.	Nearly 55,000 were signed in FY 2009, up from 20,600 in FY 2007. See ESGR, undated (f).
Patriot Award	Employees, families, and Reserve and Guard personnel can nominate employers online. Every employer nominated receives an award. The nomination is the first step in the award process.	In FY 2011, 16,560 Patriot Awards were presented. See ESGR, undated (b).
Above and Beyond Award	State and local ESGR committees give this "second-step" award. It recognizes employers that have gone above and beyond the legal requirements in support of military duty.	See ESGR, undated (a).
Pro Patria Award	One Pro Patria Award is presented to one small, one large, and one public-sector employer in each state that has provided the most-exceptional support of U.S. national defense through leadership practices and personnel policies.	See ESGR, undated (e).
Secretary of Defense Employer Support Freedom Award	The Secretary of Defense presents this award annually to the nation's most-supportive employers. The ceremonies are held annually at the Pentagon, where up to 15 employers are recognized at a time.	This award program started in 1996; since that year, more than 160 employers have received this award. See ESGR, undated (f).
Ombudsman	ESGR provides free mediation services to service members and their employers. Inquiries are handled through ESGR's Customer Service Center in Arlington, Virginia. Cases are assigned to 800 trained ombudsmen throughout the country.	The number of USERRA inquiries increased from 13,116 calls in FY 2007 to 15,870 calls in FY 2009. However, the number of USERRA cases mediated by ESGR ombudsmen has remained consistent (approximately 2,500 per year). The average number of days it takes to mediate an assigned case is ten. DOL investigations and U.S. Department of Justice (DOJ) court cases have shown similar trends. There were six DOJ cases in FY 2007 and 22 in FY 2009. See ESGR, undated (f).
Extraordinary Employer Support Award	This award was created to recognize sustained employer support of Guard and Reserve service members. Only prior recipients of the Freedom Award or the Pro Patria Award are eligible for consideration by the state committee.	See ESGR, undated (c).

NOTE: SOS = Statement of Support.

programs and whether additional programs, such as those used in other countries, should be considered.

Employer Support Programs Used in Other Countries

To inform policy recommendations, we reviewed literature on comparable programs and policy in other major developed countries—specifically, Australia and the United Kingdom.

In contrast to the United States, the United Kingdom and Australia do offer compensation programs for employers of reservists (in addition to reservist job protection legislation) in the event of an activation (Busby, 2010).

In the event of extreme hardship, there is a process through which UK employers can appeal to defer or revoke a reservist's activation. Communication between the services and employers is an essential part of this process. As of 2004, the UK Ministry of Defence (MOD) requires all reservists to provide employer contact information to the Employer Notification (EN) system and give the MOD permission to contact their employer.[5] The employer contact information is used to support a comprehensive system of employer notification. A UK reserve forces member's commanding officer (or head of branch) writes directly to the member's civilian employer after the member has served in the reserves for four weeks. The employer is notified of the benefits, rights, and obligations associated with employing a member of the reserve forces. Employers are provided with information about reservists' expected training activities, as well as the likelihood that they will be called up (activated) for a longer period of time.

Subsequently, when reservists are being considered for deployment, their employer details are brought up to date and a call-up notice is sent to the employer. The employer is given seven days from the time the notice is served to file an application with an adjudication officer (AO) for the notice to be deferred or revoked. In the application, the employer is able to describe the ways in which the business will be harmed by the loss of the reserve member employee. The definition of *harm* is imprecise; it can include anything from reputational harm to the impairment of producing new products and services. It is up to the employer to make the best case possible for why the employee should not leave his or her place of employment. To make the best decision possible, the AO will then compare the harm done to the business with the operational need for the reservist. The AO can uphold, defer, or dismiss the request. If the employer is not satisfied with the AO's decision, the employer has five days to file an appeal with the Reserve Forces Appeal Tribunal for a hearing of the case.

According to MOD officials,[6] most employers do not challenge the notice, and many of those that do file an application for deferral or revocation agree to the acti-

---

[5] A process does exist for reservists to apply for a waiver of this requirement.

[6] Personal communication with Brigadier Sam Evans and Wing Commander Charles Anderson on November 6 and November 9, 2012. These individuals provided a description of and insight on policies and experiences

vation after communication with AOs. Appeals come from all types of employers, and MOD officials do not observe any clear patterns in terms of the characteristics of employers that appeal the notices.

In the United Kingdom, employers can also apply for assistance to cover one-time transition costs (e.g., hiring or training a replacement), additional salary costs up to £110 per day (e.g., to cover the salary of a replacement that exceeds that of the reservist or overtime for other workers), and retraining costs (when the reservist is rehired). Also, the MOD pays the employer contributions to pension programs while the reservist is activated (Supporting Britain's Reservists and Employers, undated).

In Australia, the Employer Support Payment Scheme provides a fixed weekly payment for up to 78 weeks to any employer that has an activated reservist (Defence Reserves Support, undated [a]). The level of payment does not vary depending on the characteristics of the employer, the reservist, or the activation (except under special circumstances with administrative approval, such as health care professionals). In 2011–2012, the weekly rate was $1,288.10. It is adjusted annually to reflect average weekly full-time earnings based on data from the Australian Bureau of Statistics. In addition, there is a process through which employers can apply for additional assistance in the event of "substantial financial hardship or loss" (Defence Reserves Support, undated [b]).

A recent study of employer perceptions of Australian Army Reservists (Orme and Kehoe, 2012) summarizes findings from a survey of 126 employers that participated in an Exercise Boss Lift (modeled after the U.S. practice). This program, sponsored by the Australian Defence Force, took employers to visit their employees overseas. Notably, a minority of the Australian employers reported a negative aspect of the deployment on their business. The Australian employers expressed a desire for better information about duty dates, a desire to receive messages from the reservist during the absence, and guidance regarding the effective management of the transition back to civilian life. It is worth noting that the Australian employers surveyed for this study are not a random sample but a group that took the time to participate in this event sponsored by the Australian Army Reserve. These employers are likely to have a more favorable view of the reserves than a regular employer might.

Both Australia and the United Kingdom allow self-employed individuals to access these supports.

Summary

This chapter summarized a large number of factors that could plausibly influence whether an employer experiences a duty-related absence and the implications of that

related to the deployment of reserve forces and the employer appeal process.

absence if it occurs. The objective of our inquiry into the cost of duty-related absences is to explore whether any single factor or set of factors has a systematic or significant influence on costs such that those factors should be addressed through a revision to policy, practices, or programs. For example, are federal employers, small employers, or first-responder employers more likely to experience a cost impact? Would shorter absences or less frequent absences or more advance notification lessen the cost implications for employers? Are there systematic policy changes that would benefit employers across the board?

In the next chapter, we examine the available evidence on the relationships between the factors described here and employer impact and how features of USERRA, DoD policy, and ESGR programs influence that impact.

CHAPTER FOUR

USERRA Protections: Employer Impact

USERRA provides reemployment rights to individuals who are absent from their civilian employers because of military duty. As noted in Chapter Two, when passing USERRA, Congress anticipated that the reemployment provisions could impose costs on employers, but it deemed those to be acceptable in view of the benefits of the law.

As discussed in the previous chapter, there is empirical evidence or a theoretical basis to believe that characteristics of the absence, the employer, and the RC member or the position he or she holds influence the cost of a duty-related absence. In this chapter, we explore whether there is a single factor or set of factors that appears to have a systematic or significant influence on employers. Our multimethod study design allows us to tap several sources of data to answer that question. Analysis of data from SOFS-Rs, nationally representative surveys of RC members, allow us to characterize the employers of RC members, examine the frequency of absences and whether certain types of employers are more likely than others to experience an absence, examine the frequency with which RC members experience job changes or USERRA issues following duty-related absences and whether that frequency is related to employer characteristics, and explore RC members' views regarding how supportive their employers are and whether those views are related to employer characteristics. Analysis of data from the DoD National Survey of Employers provides a perspective from a nationally representative survey of employers regarding their experiences with duty-related absences, their levels of support for reserve duty, their perspectives on factors that make absences costly, and what DoD could do to ameliorate the impact. Open-ended survey responses and data from employer interviews provide concrete examples and contextual information and highlight issues that were either not addressed in the survey or were complex and not easily covered by the survey. Interviews with RC chiefs provide some additional context.

We begin this chapter with a description of the characteristics of RC employers. We then discuss how supportive civilian employers are of RC members, drawing on data from both employers and RC members. Next, we describe the prevalence of USERRA problems as reported by RC members and how those problems are related to characteristics of the employer. We then describe what employers have to say about the costs of duty-related absences, drawing on data from the National Survey of Employers and employer interviews. We conclude with a discussion of problems experienced by RC members when they return from duty-related absences. Although this report is focused primarily on the reemployment provisions of USERRA, we briefly discuss the potential implications of protections against discrimination.

Characteristics of Reserve Component Members' Employers

USERRA protections against discrimination in hiring benefit all current and former service members. The reemployment protections benefit service members who are employed at the time their military duty begins and would like to return to the same employer after that duty ends.

We analyzed data from the 2011 SOFS-R to characterize the employers of RC members.

Most Reserve Component Members Work for Large Employers
Most RC members who are not activated are employed full time in the civilian workforce. Of those who were not activated at the time of the survey in January 2011, 76 percent reported having a full-time (at least 35 hours per week) civilian job. Of these, 92 percent reported working for an employer (in contrast to being self-employed or working in a family business). Figure 4.1 shows where RC members with full-time

Figure 4.1
Civilian Employment of Reserve Component Members Employed Full Time, January 2011

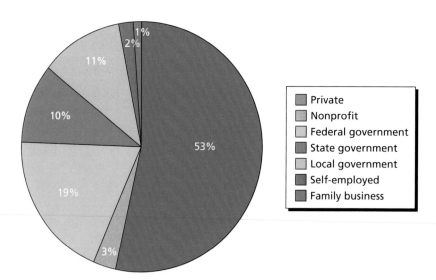

SOURCE: Authors' analysis of 2011 SOFS-R data.
RAND RR152-4.1

jobs are employed. The private sector is the most common employer of RC members, followed by federal, state, and local governments. Eight percent of the survey respondents reported being self-employed or employed in a family business. In subsequent analyses, these RC members are not considered to be working for employers.

As illustrated in Figures 4.2 and 4.3, a majority of employed RC members in both the private and public sectors work for employers with 500 or more employees, although the share working for these largest employers is larger in the public sector.[1] Fourteen percent of RC members employed in both the public and private sectors work for organizations with 100 to 499 employees. Twenty-eight percent of those employed in the private sector work for employers with fewer than 100 employees total at all locations in the United States, and 9 percent work in very small businesses of one to nine employees. For those in the public sector, only 11 percent work at businesses with fewer than 100 employees.

A Disproportionate Number of Reserve Component Members Work in Federal Agencies

In view of the increased focus on USERRA within the federal government, findings from our analyses focused on federal employers are of interest. When evaluat-

Figure 4.2
Size Distribution of Private-Sector Civilian Employers of Reserve
Component Members Employed Full Time, January 2011

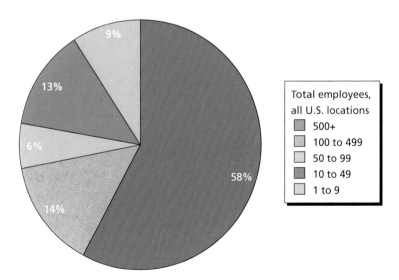

Total employees, all U.S. locations
- 500+
- 100 to 499
- 50 to 99
- 10 to 49
- 1 to 9

SOURCE: Authors' analysis of 2011 SOFS-R data.
RAND RR152-4.2

---

[1] This survey asked about the number of employees overall rather than at the specific work location of the RC member.

Figure 4.3
Size Distribution of Public-Sector Civilian Employers of Reserve
Component Members Employed Full Time, January 2011

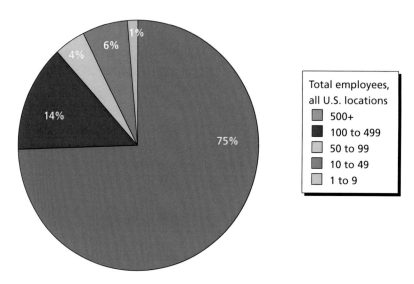

SOURCE: Authors' analysis of 2011 SOFS-R data.
RAND RR152-4.3

ing such broad and anecdotal assessments of the federal government's implementation of USERRA, it is important to keep in mind that the federal government employs a disproportionately large fraction of RC members. Whereas the federal government employs less than 2 percent of the U.S. workforce (Falk, 2012), the 2011 RC member survey reveals that the federal government employs 19 percent of the RC members who are employed full time (see Table C.7 in Appendix C). As we describe in the sections that follow, our analysis reveals no evidence that federal agencies are less supportive of RC member employees than other employers are.

Employer Attitudes Toward Reserve Component Members and Their Military Duty

Responses to the National Survey of Employers reveal that, overall, employers are satisfied with RC members as employees, view RC members favorably, and report that the military training is a benefit to their business. As reflected in Table 4.1, a strong majority of RC employers agree or strongly agree with statements related to the training and overall performance of RC employees in their organizations. Fewer than 4 percent disagree or strongly disagree with these statements. Table 4.2 shows that RC employers rate RC members as equal to or better than other employees on some presumably desirable employee characteristics (organizational skills, communication skills, man-

Table 4.1
Employer Perspectives on Employing Reserve Component Members, All Employers (%)

Perspective	Strongly Disagree	Disagree	Neither Agree Nor Disagree	Agree	Strongly Agree
The training and experience received by a National Guard or Reserve member makes that person a more valuable employee for my business.	1	2	27	37	34
National Guard and Reserve employees in my business are good team players.	1	2	18	42	37
Overall, I am satisfied with National Guard and Reserve employees in my business.	1	1	12	46	40
Employing National Guard and Reserve employees is challenging because of their military obligations.	7	19	33	29	11

SOURCE: Authors' analysis of 2011 DoD National Survey of Employers data.

NOTE: Margin of error ranges from ±0.2 percent to ±1.9 percent. Weighted Ns: more valuable = 122,003; good team players = 121,851; overall satisfaction = 122,028; challenging = 121,739. Unweighted Ns: more valuable = 7,193; good team players = 7,177; overall satisfaction = 7,188; challenging = 7,172.

Table 4.2
Employer Perspectives on Reserve Component Employees Relative to Other Employees, Reserve Component Employers (%)

Perspective	A Lot Worse	Worse	About the Same	Better	A Lot Better
Organizational skills	0	1	46	38	15
Communication skills	0	2	51	34	13
Management skills	0	2	53	31	13
Technical skills	0	1	51	35	13
Leadership skills	0	2	41	39	19
Teamwork skills	1	2	33	43	21
Dependability	1	2	36	36	25
Initiative	0	2	44	35	18

SOURCE: RAND Analysis of 2011 DoD National Survey of Employers data.

NOTE: Margin of error ranges from ±0.1 percent to ±2.0 percent. Weighted Ns: organizational = 120,846; communication = 120,504; management = 120,550; technical = 120,430; leadership = 120,687; teamwork = 120,693; dependability = 120,920; initiative = 120,546. Unweighted Ns: organizational = 7,114; communication = 7,091; management = 7,084; technical = 7,091; leadership = 7,099; teamwork = 7,104; dependability = 7,114; initiative = 7,102.

agement skills, technical skills, leadership skills, teamwork skills, dependability, and initiative). A very small share of employers rated RC members as worse than other employees on these dimensions. Analyzing relative ratings of RC members, we find that larger firms were less likely to view RC employees favorably than non-RC employees (see Appendix B, Table B.26 and related discussion).[2] We also find that federal employers were more likely than private employers to view RC employees favorably compared with non-RC employees.

When asked, nearly all employers interviewed for this study—regardless of size—reported that RC employees are an asset *when present*, and they specifically cited discipline and reliability as favorable traits common among RC members. Employers described RC members as being "top notch," "professional," and "high caliber." On the other hand, one large employer suggested that—although RC employees are viewed as an asset—an RC member's military background and "regimented" personality might not lend themselves to working in the retail industry, in which "every day is different."

The high level of support expressed by employers in the survey is consistent with what RC members have to say about how supportive their employers are of their reserve duty. In the 2011 SOFS-R, few RC members reported that their employers were unsupportive of their service (see Tables C.13–C.14 in Appendix C).[3] Overall, 5 percent reported that their employers were unsupportive, and 2 percent reported that their employers were very unsupportive of their service. However, those members who had not been activated in the past 24 months were more likely to report that their employers were very supportive (see Appendix C, Table C.16 and related discussion), suggesting that activations do have an impact on how service members view their employers' support. Examining employer support by sector, we see that RC members who are employed by the federal government are more likely to describe their employers as very supportive. RC members employed by larger employers (500 or more employees) were most likely to describe their employers as very supportive (see Appendix C, Table C.14).

USERRA Problems Reported by Reserve Component Members

The 2011 RC member survey also sheds light on the prevalence of USERRA problems. Among all RC members who reported working in the private, public, or nonprofit sector prior to their most recent activations, 18 percent reported experiencing a USERRA problem since returning to their civilian jobs. There has been a decline in reported USERRA problems since June 2008, when 22 percent of RC members

---

[2]   We averaged the responses to all the comparison items (assigning a point value of 1–5 for each response, where 1 was assigned to "a lot worse" and 5 to "a lot better").

[3]   The survey asked the RC member about the degree of support of both the immediate supervisor and the employer. We report the responses to the question about the employer. Responses to the two questions were generally similar, but a significant number of RC members reported that they did not have a supervisor, so there were fewer responses to those questions.

reported USERRA problems. Among those who reported any USERRA problem, a small fraction reported that the employers could not accommodate reemployment claims or that they were terminated without cause after deactivation.

We explored the likelihood that an RC member would report any USERRA problem related to the size and sector of a firm, controlling for whether the respondent returned to the same employer. None of the characteristics (including whether the employer was a federal agency) was statistically significant (see Appendix C, Table C.44).

Overall, RC member survey responses indicate that most RC members feel that their employers are supportive, relatively few experience USERRA issues, and few of those who experience USERRA issues are terminated or not reinstated to their jobs. There is a fair amount of churning that coincides with an RC member's return from duty, and it appears that much of it is due to RC members' choices. Some RC members do not return to their predeployment jobs because of changed circumstances for their employers (layoffs, out of business). However, our analysis does suggest that RC members are more likely to view their employers as very unsupportive when they have experienced a recent deployment. This raises the possibility that employers that are otherwise supportive of military duty become unsupportive when faced with the realities of activation.

When asked how supportive their employers are of their military service, RC members working for the federal government were the most likely to report "supportive" or "very supportive." A regression analysis revealed no statistically significant differences in RC member reports of employer support between federal and private sectors (Table C.17) after controlling for other factors.

In spite of this overall favorable impression, more than one-third of employers that were surveyed as part of the National Survey of Employers agreed or strongly agreed that employing RC members is challenging because of their military obligations (Table 4.1), and a majority of employers indicated that a one-year duty-related absence every five years would hurt their business (Figure 4.4).

Business Challenges

Unlike many other employment laws (e.g., the FMLA) that impose recordkeeping requirements that may be burdensome to employers, the employer impact of USERRA stems primarily from the effects of dealing with a duty-related absence.

Few U.S. Employers Experience Duty-Related Absences

According to the sampling frame used by DMDC for the National Survey of Employers, the number of unique CEI employers (those officially documented as employing RC members in the CEI database) account for less than 2 percent of all

Figure 4.4
Impact on Business if Reserve Component Member Were Absent One Year of Every Five

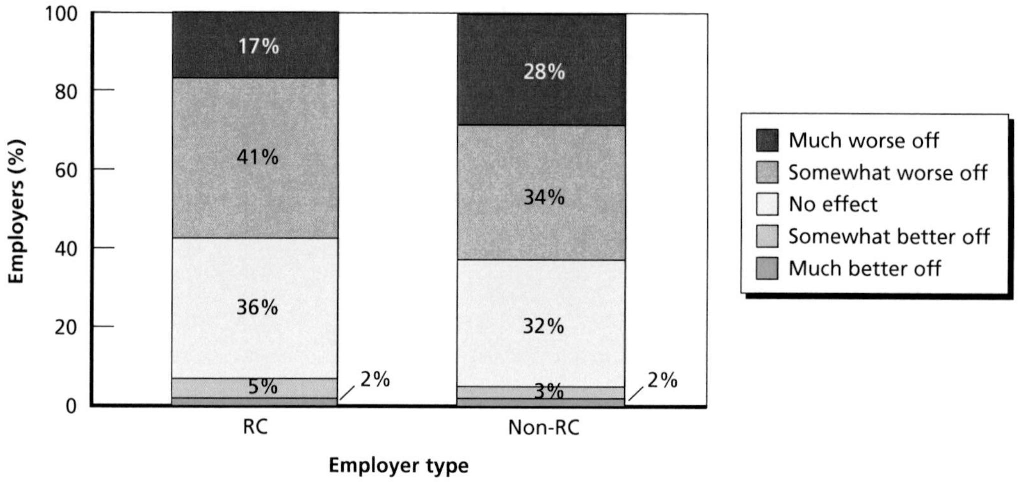

SOURCE: Authors' analysis of 2011 DoD National Survey of Employers data.
NOTE: Unweighted N for RC = 9,530; for non-RC = 869.
RAND *RR152-4.4*

U.S. employers. Among CEI employers that responded to that survey, 26 percent reported employing no RC members in the previous 36 months.[4] Approximately one-third reported employing one RC member, while another third reported employing between two and ten RC members. Only 7 percent of the CEI sample reported employing 11 or more RC members. Thus, although activation is a common occurrence for RC members, with about one-third of RC members reporting having been activated in the previous two years as of January 2011, it is rare for a "typical" U.S. employer to experience a military duty–related absence.

When an Absence Occurs, the Impact Varies

Echoing themes from the prior literature, employers report that business challenges resulting from duty-related absence are real but varied. Although we do identify some systematic relationships between employer characteristics and business impact, interviews with employers and open-ended survey responses highlight the fact that employer impact will depend on the characteristics of the employee (including how the employee handles notification), the characteristics of the absence, and the employer's

---

[4] There are many reasons that survey respondents from the CEI sample may have reported that they had not employed an RC member in the previous three years. First, the CEI database used for sampling purposes was from December 2009, whereas the employer survey was fielded in 2011. Second, an RC member could register his or her employer but not notify the employer of his or her RC status. Third, the individual responding to the survey for the employer may not have been aware of employees' RC status.

circumstances at the time of the activation. Many employers struggled to describe the impact of "an activation" on their businesses in the National Survey of Employers, as the following comments indicate:

> Some of the questions offered only yes or no answers when my answer would have been maybe. In my business, answering these questions for some class of employees would be different than for others. For example, we have only one office manager, only one billing clerk, and only one receptionist—if any of them were Guard or Reserve their absence would be much harder to deal with than say a secretary, paralegal or lawyer of which we have many and thus could "cover" their absence more easily. These questions don't allow for those differences. (Survey responses, question 47, CEI firm, 11–49 employees)

> The questions that ask to reply about "an employee" are vague. There are some employees that would be very difficult to replace and there are others that could be replaced in a day. I responded considering an average employee but I think that the feedback you receive would be more relevant if you had specified "think about an employee in a critical position" or "think about your average employee," etc. (Survey responses, question 47, CEI firm, 50–99 employees)

Most Employers Report No Business Challenges Resulting from Duty-Related Absences

Analysis of data from the National Survey of Employers reveals that, among employers that reported the activation of at least one RC employee in the previous three years, 28 percent reported that the absence resulted in a change to business operations. Public employers (both FR and non-FR) were more likely than private employers to report changes to standard business operations. A larger share of employers that reported no business impact had not experienced absences longer than 30 days (32 percent) than of employers that did report a business impact (22 percent), suggesting that absences longer than a month pose greater challenges for employers. Still, it is worth noting that more than 30 percent of employers that reported no business impact had experienced absences of longer than one year. Employers with one to ten employees were more likely to report changes than all other size categories of employers, with the difference between these and larger firms being greater as the size category increases (see Appendix B, Table B.10). Overall, the survey responses suggest that about one-quarter of employers report business changes due to absences and that public employers and smaller employers are more likely to report such changes.

How Military Duty Affects Employers

Respondents to the National Survey of Employers who did report that military duty affected their business operations were asked to describe the extent to which they experienced each of 12 challenges. For analytical purposes, we grouped 11 of these challenges into three categories based on a factor analysis of survey responses, as follows:

- managing workload
  - increased workload of coworkers
  - disruptions in work scheduling
  - disruptions in product delivery or workflow
  - lower coworker morale
  - loss of critical work skills
  - increased overtime costs
- replacement workers
  - increased costs from hiring replacements
  - increased costs from training replacements
  - increased time spent finding or training qualified replacements
- loss of business
  - loss of existing business
  - difficulty developing new business
- not categorized
  - increased cost of benefit plans
  - other.

Figure 4.5 reflects employer ratings of the extent to which they experienced each challenge. The pattern of responses reflects the fact that employer impact varies by employer. Each challenge was reported by at least some employers, and, at the same time, some employers reported that those challenges were not an issue at all.

Overall, challenges related to managing workload are the most common and significant. Within that category, lower coworker morale was the least common issue. Loss of business was highlighted as a moderate, large, or very large issue by less than 20 percent of employers. However, employers with one to ten employees were more likely to experience client problems than all other size categories of employers, with the difference between these and larger firms being greater as the size category increases (see Appendix B, Table B.12). An analysis of the survey data also suggests that employer characteristics were not related to whether a business experienced issues related to managing its workload. Private non-FR employers were more likely to report challenges related to replacement workers than were public non-FR employers.

A separate survey question asked the same subset of employers what methods were used to adapt to duty-related absences. When asked about methods used to adapt to RC member absences, only 11 percent of employers reported suspending or delaying

Figure 4.5
Extent of Business Challenges Experienced During Employee Leave

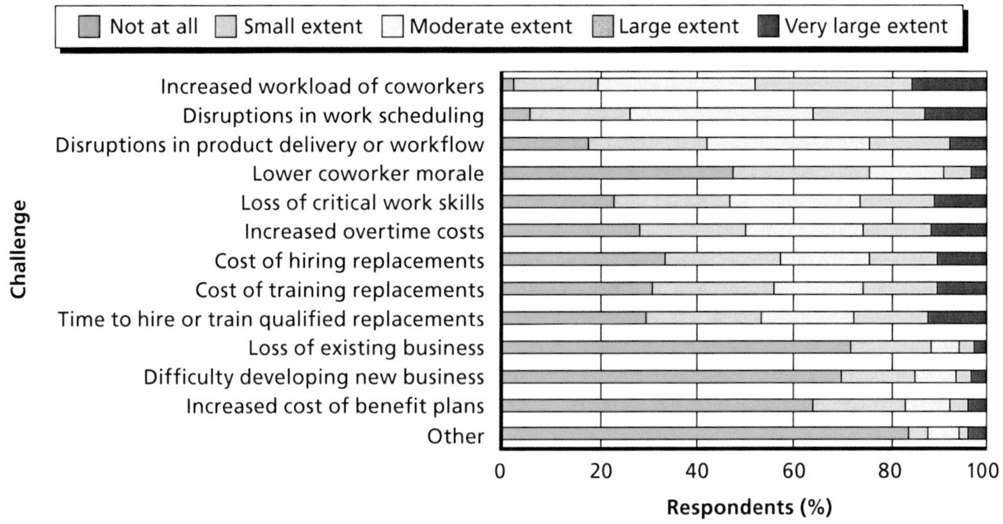

SOURCE: Authors' analysis of 2011 DoD National Survey of Employers data.
RAND RR152-4.5

business; 12 percent reported allowing work to build up until the employee returned (see Appendix B, Table B.15). The most common method was dividing responsibilities among coworkers, with 86 percent reporting this. Reassignment of responsibilities to a single coworker (52 percent) and supervisors assuming responsibilities (59 percent) were also common methods. Less than half of employers reported hiring either temporary or permanent replacements (39 percent and 24 percent, respectively). An analysis of the survey data reveals that it is the smallest employers (those with one to ten employees) that were more likely to allow work to build up or suspend business operations during a duty-related absence. These employers were also more likely to shift responsibilities to a single coworker or a supervisor and hire temporary replacement workers (see Appendix B, Table B.16 and related text). Public (non-FR) employers were less likely to hire replacement workers, and public (FR and non-FR) employers were also less likely to suspend business operations.

Our interviews and analysis of open-ended survey responses suggest that public employers are often restricted from hiring replacement workers and are instead required to offload the work of an absent worker to other current employees or hire contractors to accomplish the work. One municipality reported that "union issues" arise if circumstances dictate that an employee is promoted to fill a gap and that the promoted employee must subsequently be demoted when the RC member reassumes his or her position. Another public employer pointed out similar difficulties, noting that the city would not hire a replacement for RC members because (1) RC members continue to accrue benefits during their absence and the city compensates for pay differential and

(2) policy prohibits offering the position and corresponding duties to a new, external hire who would vacate the position upon the RC member's return.

Because survey questions asked about the cost and time involved in hiring replacement workers, the fact that some public employers are restricted in their ability to hire replacements may explain why public employers were less able to use this method to adapt and were less likely to report problems related to replacement workers. Less than 20 percent of employers reported being affected to a moderate, large, or very large extent by the increased cost of benefit plans. However, an analysis of survey responses on this item revealed that firms with 50–99 employees were significantly less likely to experience increased benefit costs than firms with one to ten employees (Appendix B, Table B.12). For example, one public employer stated its policy of continuing to provide health benefits—regardless of deployment duration—at an average cost of $28,000 per year per family. A representative from another public employer bemoaned the strain on the city budget caused by the continuation of benefits for absent employees, adding that pension payments were the most detrimental.

The interviews and open-ended survey responses raise several employer challenges that were not fully captured by survey items. Some employers in school systems provided write-in survey responses emphasizing the adverse effect that deployment has on students when a teacher or principal is absent for military duty:

> My concern as an office manager, is that many of our children experience anxiety whenever their teacher is gone for just one day, and when we have a teacher on six weeks' maternity leave, the children become even more anxiety-ridden and our work load rises. The best for our children is to have every staff member consistently available for the majority of the school year with other responsibilities perhaps scheduled in the three months of summer when the children have gone home. I have family members who are members of the National Guard so I am familiar with how great of a need we have for them in our world and the terrific job they do. Thanks for everything you do. (Open-ended survey responses, question 47, CEI firm, 100–499 employees)

> As a public school, having one of our teachers gone for any length of time is not only a financial burden to our district but also a disservice to our students. Long-term subs are not always available in the area [in which] the member absent is certified. (Open-ended survey responses, question 19, CEI firm, 50–99 employees)

The challenges expressed by those in school systems were echoed by other employers in client-oriented service businesses in which employees develop relationships with clients and in which those relationships are important to service provisions; examples include mental health services, accounting, and consulting. Employers generally emphasized that, with such employees, it is not possible to simply transfer the work to another person without a loss of service quality and that, in some cases, the member's absence can lead to a loss of business. An interviewee representing a nonprofit serving abused and neglected youth (a large firm employing approximately 400 employees) described the employer's past experience with an RC member's absence and lamented the deployment's potential effect on those involved in the program.

Some employers, particularly those in the public sector, cited lost productivity as an important issue. One employer we interviewed (a municipality with approximately 250 on payroll) complained that, because the employer is explicitly precluded from hiring a replacement worker for someone who is on leave, productivity suffers and there is work that simply does not get done. This is not a loss of business but rather a loss of productivity. Another interviewee, an automotive retail firm (employing just over 100 workers) discussed how a loss of productivity resulting from an RC member's absence could translate into a loss of business. If productivity suffers and there are delays in servicing vehicles, the service department might see a decline in patronage.

Characteristics of the Military Absence and Position of Military Member Influence Effects

Respondents to the National Survey of Employers who reported that military duty affected their business operations were asked to report on the extent to which each of eight aspects of military absence contributed to business challenges. Figure 4.6 reports how employers rated the significance of each aspect of the military duty in terms of the extent to which it contributed to business problems. Overall, the length and frequency of absences were cited most frequently and reported to pose the most-significant challenges. It is notable that each aspect was reported by at least some employers to contribute to business problems to a very large extent. An analysis of the relationship between employer characteristics and whether the feature of the absence posed an issue for the

Figure 4.6
Employers Reporting Feature of Military Absence Contributed to Business Problems

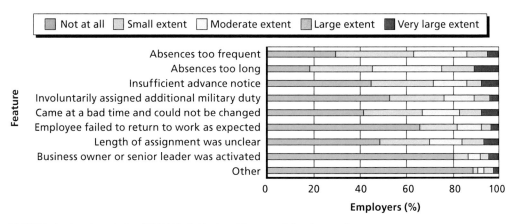

SOURCE: Authors' analysis of 2011 DoD National Survey of Employers data.
RAND RR152-4.6

employer revealed that public organizations felt that there are multiple aspects of military absences that contribute to their business problems. It also revealed that smaller firms (those with one to ten employees) reported issues with frequent absences and inconvenient call-up timing (see Appendix B, Table B.14 and related text).

Exploring the responses for the "other" option for this survey question, we saw multiple comments regarding the absence of key employees presenting additional hardships. It could be that, to mitigate hardship when those employees are activated, some firms are not promoting RC members to key roles. One of the public employers that we interviewed (a public utility firm with more than 500 employees) surmised that RC members generally seek positions with duties that can be performed by colleagues during absences—which explains the dearth of RC employees in senior management positions. An automotive retail employer we interviewed (employing just over 100 workers) stated that frequent absences, often of indeterminate duration, affect an RC employee's ability to gain experience, skills, and expertise that might render him or her suitable for a higher-salaried position.

Some firms were unclear as to whether RC absences are voluntary or mandatory. Additionally, two firms expressed suspicion in open-ended survey responses that some employees had used their military duties to escape work obligations. Other firms mentioned that a lack of knowledge regarding the timing of employee return presents hardships (presumably in terms of planning and hiring a replacement); lack of advance notice was again mentioned:

> We have no idea when our employee is returning. (Open-ended survey responses, question 13, CEI firm, 11–49 employees)

> It's never made clear whether or not the absences are voluntary or mandatory. It would be nice to know that information so I could make future business decisions. I.E whether or not to hire a Guard or Reserve member. I understand mandatory obligations but don't understand why voluntary assignments should affect my business. (Open-ended survey responses, question 13, CEI firm, 100–499 employees)

> It would seem that National Guard employees are best suited to entry-level or line positions where their absence would not be as disruptive as if they are in management or key positions. (Open-ended survey responses, question 13, CEI firm, 11–49 employees)

In interviews and open-ended survey responses, employers emphasized challenges associated with military duty of a highly skilled or trained employee or an employee with special certifications. One interview with a law enforcement agency (a large public employer with more than 2,000 employees) revealed that, due to the small proportion of the employee base they constitute, canine handlers and motorcycle patrol officers are incredibly difficult to replace. The employer stated, for example, that the department employed fewer than 30 canine handlers among more than 1,000 sworn officers. The same interviewee went on to describe the difficulty posed by the long-term deployment of one assistant commander, in which the burdened district commander was limited to seeking employees in equal-rank positions to fill in during the RC member's absence. An interviewee from a large public utility company (employing more than 500 employees) discussed the inconvenience of having an RC member who has been employed for six months, is then deployed while still in training for the job, returns six months later after deployment, and has to start training over.

A related challenge that employers face is dealing with the absence of "the only" employee of a certain type in an organization. Although such an employee might not be a senior leader or even highly skilled, his or her absence can represent a big loss for the employer. This is especially true for public employers that are not allowed to hire replacement workers. One employer interviewed for this study (a municipal government employing approximately 250 employees) mentioned that it is relatively easy for the city to deal with the absence of a firefighter because there are 20 other firefighters to whom the work can be offloaded. However, the interviewee was not sure what the city would do in the case of an extended absence of a person filling a unique position (such as the grant writer). If the job is being held for the individual, it cannot be filled by a replacement. So the city would need to shift the work to other people if possible (who may not have the skills) or hire a contractor to do the work. Higher-education institutions face a similar problem, according to our interviews. One large public employer in the higher-education field (employing more than 500 employees) pointed to a state policy affirming that, if a deployed RC employee were a staff member, he or she could be replaced by a temporary employee. However, the absence of a faculty member would likely pose a greater challenge for an affected department at the university.

Experiences of Reserve Component Members and Employers After Duty-Related Absences

USERRA requires employers to reemploy RC members after a duty-related absence. However, not all RC members do, in fact, return to the same employer. In this section, we present information on the postactivation employment experiences of RC members and their employers.

Reserve Component Members May Not Return to the Same Employers After Duty-Related Absences

The 2011 SOFS-R data indicate that most, but by no means all, RC members who are activated return to the same civilian employers after their duty ends. A key USERRA-related concern is whether those who change employers do so by choice or because they were denied reemployment.

RC members who were activated at the time of the survey were asked about their postdeployment employment intentions. Many of the surveyed RC members who were employed prior to activation reported that they did not plan to return to the same employers after deactivation. Among those who had been employed,[5] 61 percent reported a plan to return to the same employer. Among the self-employed and those employed in a family business, only 55 percent reported such a plan.

RC members who were not activated at the time of the survey but had been activated since September 11, 2001, were asked about their current employment status, as well as their employment status prior to their most recent activation and in the three months following their most recent deactivation. Sixty-one percent of those activated since 2001 reported that their current civilian employers were the same as prior to their most recent activation; 29 percent had different employers prior to activation, and 10 percent did not have civilian jobs prior to their most recent activation.

Fifty-nine percent reported working for pay or for profit in the three months after deactivation. Of those reporting private, public, or nonprofit employment both before and after activation, 87 percent said that they had returned to the same employers, while 14 percent had changed employers. Of those who changed employers, a majority (55 percent) reported that they had found better jobs.[6] Thirty-five percent reported that the decision not to return to their preactivation employers was influenced by changes to the employers' circumstances during deployment (including business closure, layoffs, and lack of prompt reemployment). Four percent reported that they did not receive prompt reemployment but did not cite any particular issue with the predeployment employers' business. Five percent reported that they were recuperating from illness or injury.

Those who did not return to work in the three months following deactivation were asked about the factor that contributed to their nonworking status. Among this group, 42 percent reported that they needed a break, 25 percent decided to attend school, and 12 percent reported that they were recuperating from illness or injury. Thirty-three percent reported that the decision not to return to work was influenced by changes to the employers' circumstances during deployment (including business closure, layoffs, and lack of prompt reemployment). Three percent reported that they did not receive prompt reemployment but did not cite any particular issue with the predeployment employer's business.

Our analysis of the survey data also reveals that, compared with those currently employed in state and local government, federal employees are much more likely to

---

[5]   This excludes individuals who were self-employed or employed in a family business.

[6]   Survey respondents were allowed to respond "yes" or "no" to each of seven possible factors, and we cannot isolate what may have been the most significant factor.

have been either not employed or working elsewhere prior to their most recent deployment (see Table C.9 in Appendix C). This suggests that RC members may be particularly attracted to federal employment. Survey responses suggest that federal employers are more likely to provide a continuation of full pay and health benefits than employers in other sectors are (see Tables C.4 and C.5), which may be a reason that RC members are attracted to federal employment.

Analysis of responses to the RC member survey reveals that activation and deactivation are associated with a fair amount of employment turnover, much of which is voluntary. Very few RC members reported a lack of prompt employment that is not associated with changes to the preactivation employer's circumstances.

Challenges Faced by Employers When Reserve Component Members Return from Duty

Although the survey emphasized challenges that an employer faces while an RC member is absent because of military duty, interviews and survey responses also highlight some challenges involved in dealing with the RC member's return. One large public-sector employer we interviewed told of one RC faculty member who was deployed in the middle of the academic year. Said employee reassumed his position upon return from deployment—but with an alternative assignment because his replacement hire was still (presumably under contract) with the university. One small employer we interviewed estimated that RC members (mostly USAR or ARNG) constituted anywhere from 5 to 8 percent of its workforce. The participant spoke about an experience with a returning RC employee whose mental affect had changed so much that he was no longer suitable for security work; the employer also indicated that physical injuries might preclude a returning RC member from reassuming his or her position.

Multiple firms mentioned that, when some RC members returned to work, there were problems with readjusting to civilian life, particularly when an employee was injured. One employer representing the public sector described an incredibly comprehensive, proactive approach to employee reintegration in the law enforcement field. Spearheaded by the department's superintendent—who, concurrently, is a licensed psychotherapist—the process relies on contracted department psychologists to engage returning employees in combat-veteran discussion groups, as well as conduct one-on-one sessions. The department also establishes "family liaisons" to facilitate an uninterrupted relationship with RC employees' family members. The employer in question reported having consequently received various ESGR awards and recognitions.

Conversely, our interview with one large nonprofit employer not only revealed overwhelmingly positive experiences upon RC member return but also stated that some RC employees actually had better dispositions upon returning to the workplace. Responses to the National Survey of Employers indicate that 15 percent of employers that had employees absent due to military duty in the previous three years reported that the RC member returned to work with *at least one* of the following problems:

- substance abuse that interferes with performance (e.g., alcohol, drugs)
- increased difficulty interacting with customers or coworkers (e.g., easier to anger, less helpful)
- increased risk-taking at work (e.g., less likely to follow safety precautions)
- a military service–related disability requiring changes to employee workstations, tasks, or routines
- increased stress or emotional problems.

Twelve percent of employers reported that employees returned with increased stress or emotional problems. Eight percent reported difficulty dealing with customers.[7] In open-ended survey responses, employers highlighted issues related to posttraumatic stress disorder (PTSD), getting along with clients and coworkers, and stress-related issues.[8] Federal employers reported at higher rates than other employers that RC members were experiencing problems upon return. The fraction of federal employers reporting that RC members returning from activation had problems with substance abuse, a service-related disability, or increased stress was more than twice as high as for the overall survey population. Four percent of federal RC employers reported substance-abuse problems, and 1 percent reported increased risk-taking at work. Nearly 23 percent of federal RC employers reported that their returned employees had increased stress or emotional problems, while approximately 15 percent said that these employees had increased difficulty interacting with customers or coworkers. Roughly 11 percent reported that their employees had service-related disabilities that required accommodations at work.

When a highly trained, certified, or skilled employee is absent for military duty, reintegrating them can involve costs to the business, especially in the case of a long absence. One public employer interviewed—a law enforcement agency with more than 500 total employees—described a mandated 120-hour retraining for any employee whose absence totaled 180 or more days, regardless of the nature of the absence. This resulted in a one-time cost per mobilization per RC member. Another large public RC employer we interviewed pointed to requirements for returning military members and the costs of proctoring promotional exams for those who may have missed them.

Another related issue highlighted by several employers is the hardship associated with terminating the replacement worker. This was mentioned in the open-ended survey responses from smaller employers that noted that they felt bad for the replace-

---

[7]   Each employer could select more than one problem.

[8]   The June 2008 and January 2011 SOFS-Rs asked RC members who had been activated after September 2001 whether they had been wounded during any activation. In June 2008, 18 percent of those activated reported that they had been wounded during an activation. By January 2011, that figure had increased to 22 percent. Of those wounded, 35 percent reported that the injury they received during activation limited their working ability in their principal civilian employment. Twelve percent reported that they had experienced problems with their civilian employers being unable to accommodate a disability incurred during military service.

ment worker and that they were also faced with the burden of the unemployment insurance costs associated with that termination:

> If I know I am going to be without my Reserve member for an extended period of time, I will need to hire and train a replacement for the duration of his absence. There is a substantial cost involved in doing so, and then this temp will be laid off and consequently collecting unemployment after our Reserve member returns. This is obviously not an ideal situation. Our country's freedom is anything but free and I will do my part as an employer and American. I think it would be very beneficial if I as well as other small businesses could hire temporary employees without the burden of the cost of unemployment. (Open-ended survey responses, question 47, CEI firm, one to ten employees)

> If an employee is hired to fill [an RC member's] position, that employee would have to be laid off in order to make room for the Reserve member. In addition to the discomfort of this, we are at risk of being charged unemployment tax benefits, and layoffs increase our tax rate. We shouldn't suffer a financial burden for having hired a National Guard or Reserve member who later serves an extended military tour/ duty. (Open-ended survey responses, question 47, CEI firm, 11–49 employees)

It is particularly worth noting that a handful of employers expressed a sentiment that the challenges they face in dealing with military duty are so significant that the burden has led them to avoid hiring RC members. Although these respondents were small in number, these comments from the survey responses are notable in view of USERRA provisions that prevent employers from discriminating in hiring on the basis of military service:

> What ever happened to one-year deployments every five years? How can you expect me to run a business when my employees are leaving every three and a half years for 16 months? I will NO LONGER hire National Guard members as employees. (Open-ended survey responses, question 19, CEI firm, 11–49 employees)

> My business is very small. When a key employee became a Reserve member I did not understand he could be taken away from the business for a year. I require highly trained workers who are in an unusual field, and simply cannot hire replacements quickly under any circumstances. When he was called to duty it was impossible to sufficiently train a replacement in time, so I was unable to bid contracts for my busiest time of the year. It very nearly ended my business. I support our military and the Reserve system, but (sadly) in the future I would be unable to hire Reserve members because it simply jeopardizes the very existence of my business. (Open-ended survey responses, question 19, CEI firm, 11–49 employees)

We are such a small business struggling to stay in business that at this time we are not able to support employees who need to take leave from work. We have not had a National Guard member apply, and I would inform them that we would not be able to support them in a good way. Unfortunately. (Open-ended survey responses, question 47, non-CEI firm, employer size data missing or excluded)

National Guard Soldiers are being deployed too frequently. It is coming to a point, that National Guard soldiers will not be hired in the civilian work force. (Open-ended survey responses, question 19, CEI firm, employer size data missing or excluded)

Employer Perspectives on DoD Policies and Programs

As suggested by the conceptual framework described in Chapter Three, utilization policies and practices influence the nature of activation or the types of RC members activated, which could indirectly influence the likelihood or cost of absences. Direct efforts to support employers in the event of an activation could ameliorate those costs. In this section, we describe the perspectives of employers.

No One Structure of Military Duty Is Preferred by All Employers

All employers were asked whether they would prefer shorter, more-frequent absences or longer, less frequent absences. The responses of RC and non-RC employers are summarized in Table 4.4. A majority of employers did not state a preference for or against shorter, more-frequent duty (over longer, less-frequent duty). A majority of non-RC employers reported that both duty structures would be bad for business. Employers that did state a preference for one structure over another were equally divided between

Table 4.4
Preferred Duty Structure, by Employer Type (%)

Duty Structure	RC	Non-RC
Shorter, more frequent	19	13
Longer, less frequent	18	14
Both good	29	22
Both bad	34	51

SOURCE: Authors' analysis of 2011 DoD National Survey of Employers data.

NOTE: Margin of error ranges from ±1.2 percent to ±4.5 percent. For RC, weighted N = 166,775; unweighted N = 9,521. For non-RC, weighted N = 7,051,667; unweighted N = 872.

the two. Federal employers were less likely than other employers to report that both options would be bad for business and were more likely to state a preference for shorter, more-frequent absences (see Appendix B, Table B.6).

Employers were also asked whether, in the event of a deployment to exceed one year, it would be better for the business if the two-month predeployment training occurred (1) immediately prior to the deployment or (2) one or more months prior to the deployment. Two-thirds of employers responded that they would prefer to have the training occur immediately prior to deployment. One-third of employers expressed a preference for a gap of one or more months between training and deployment. These responses suggest that an RC member's absence is burdensome no matter how it is "packaged" and that no single approach to balancing the frequency versus length of RC absences is preferred by all employers.

Employers Have Differing Perspectives on Potential Support Programs

All employers surveyed were asked which of the following hypothetical measures would be most helpful for their businesses (assuming they had an RC member as an employee):

- an incentive for hiring a National Guard or Reserve employee
- providing replacement assistance for job vacancies
- opportunity to reschedule military duty to a more manageable time
- an incentive providing partial reimbursement of employer expenses, such as a tax incentive, grant, or low-interest loan
- "None of these measures would be helpful for my business."

The most popular response was "none of these measures," with 27 percent of employers choosing this option. Twenty-five percent said that providing replacement assistance for job vacancies would be most useful. Nineteen percent would have liked an incentive to at least partially reimburse employer expenses. Sixteen percent preferred an incentive for hiring an RC member employee. Thirteen percent said that they would have liked to be able to reschedule military duty.

An analysis of the survey data did reveal some variation in desired support programs by employer characteristic. Public (non-FR) employers were less likely than private (non-FR) employers to indicate that hiring incentives or reimbursement of expenses would be helpful. The largest employers were less likely than the smallest (those employing one to ten people) to indicate that hiring incentives would be helpful. Employers with 11–49 employees and those with 50–99 employees were more likely than firms with one to ten employees (significant at the 90- and 99-percent levels, respectively) to think that rescheduling military duty would be helpful. This could perhaps indicate that, in the smallest firms, there is no "good" time for an employee to be absent (see Appendix B, Table B.20).

Employers were asked to write in suggestions for additional measures that would be helpful for their businesses. DoD recorded 4,781 responses. Many used the write-in suggestion box to emphasize and elaborate on the options provided in the survey.

Surveyed employers cited a need to plan for an employee's absence and, thus, a need to know when the individual would leave and when he or she would return. One large employer in the engineering field that we interviewed pointed out that ten to 12 weeks' advance notice for securing a temporary worker's security clearance can be a major hurdle. One automotive retail firm interviewed (employing approximately 100 people) discussed planning for an RC employee's absence, noting that a three-month absence would likely not prompt the company to hire a temporary or replacement employee but that a six- to nine-month departure likely would. Our interview with one large public employer (whose jurisdiction totaled more than 500 employees)—among others— suggests that unpredictability of deployment and return dates is a serious problem from an administrative standpoint; departments are forced to react instead of plan, affecting the city budget:

> The best help would be knowing a definite return to work date that would allow us to plan how to cover the absence in the most efficient manner. (Open-ended survey responses, question 19, CEI firm, 50–99 employees)

Of the 4,781 responses, 714 emphasized the importance of advance notice of military duty, although the desired length of notice (when mentioned) varied from one week to one year, with many simply requesting "as much advance notice as possible." One large firm we interviewed (an automotive retailer that employs approximately 100 people) distributes employee schedules in one-month intervals; thus, any employee requests for leaves or absences of any nature must be brought to the administration's attention 30 days in advance. One hundred seventy-three responses mentioned the importance of a known return date, advance notification of the RC member's return, or notification of changes to the original return date. Many of the comments on these two advance-notice topics highlighted frustration with changes to original schedules. Several employers mentioned that units send out annual drill schedules but then do not adhere to them; others highlighted deployments being canceled or time frames being shifted.

These responses regarding advance notification of the absence were often com- bined with suggestions that the notification be provided to the employer directly by DoD—ideally, in a standard format that could be understood by employers and could be used DoD-wide.

Overall, 135 respondents highlighted the need for direct notification in the open- ended survey responses. This suggestion appeared to correspond to different concerns or issues, depending on the employer. Some employers seemed to want the documen- tation for administrative or benefit-management reasons, while others wanted DoD to notify them because employees did not or could not (in the case of changes to the return date) notify the employer. Other employers indicated that they did not trust all employees and wanted the ability to verify claims of military duty. One concern was that RC members might use military duty as an excuse to get out of work or unfavorable shifts. Several employers also reported requesting documentation as to whether the duty was voluntary. A few of these responses indicated that the employers were not aware that voluntary duty is covered by USERRA.

A related but less frequently mentioned suggestion was for the military member to confirm his or her intent to return to work upon leave for duty. During our focus group session, one RC employer (a large engineering firm) suggested that military duty may lead members to change careers shortly after return but that they want the security of being employed; the employer in question had experienced resignations from employees returning from active duty. Similarly, an interviewee (from a large public employer) reported that departing RC members almost *always* declare their intent to return to work, though it had also had experiences with members who never communicated with or returned to the employer postdeployment.

During our interview, one large public employer specifically requested documentation from DoD as to whether the duty counted under the five-year USERRA limit. Representing a municipality employing more than 500 people, the employer expressed serious concerns about USERRA's five-year stipulation. Because these absences are "cumulative," the employer wondered at what point employers can terminate eligible employees. Although no one had quite hit the five-year point, the interviewee insisted, the time will come. (The employer went on to say that the threshold must be based "on WWII [World War II]," referring to its duration compared with that of the ongoing post-9/11 conflicts):

> We had an employee that was out on leave for active duty for over a year. The orders we had received indicated she had returned, but we didn't hear from her. We sent a letter to her last known address and made a phone call, but were never able to connect. From an employer standpoint, it would have been very beneficial if we had received information from ESGR or someone from the military about her return. We were a bit hesitant to make too much effort to contact because she could have had a fatal injury or other serious injury that she didn't want to make public. We finally terminated her employment after six months had passed since her expected return date. It would have helped to have had more information about her return. (Open-ended survey responses, question 47, CEI firm, 50–99 employees)

Flexible scheduling, emphasizing an opportunity for the employer to influence the timing of leave (both annual training and deployments), was mentioned in 147 open-ended responses. As noted earlier, employers representing school systems mentioned concerns regarding the deployment of teachers and its influence on students and on the school system. Among the suggestions from school systems was to schedule annual training during the summer or other school breaks and to align deployments with the school schedule to the extent possible. Respondents also suggested that schools might benefit from having teachers deployed after state tests have been given or deployed for an entire school year, rather than returning midway through the year and having to bump the replacement teacher.

Several employers, many of them small, requested that DoD limit the number of employees who could be deployed at any one time from a single firm. Employers that require workers to work weekends indicated a preference for drill duty during weekdays. Other businesses that are seasonal (e.g., accounting, agriculture, tourism) indicated that absences would be more convenient during their off-seasons.

Assistance with replacement workers was mentioned by 167 employers in open-ended survey responses. Some employers suggested that DoD should provide a replacement worker when the reservist is activated, while others were seeking assistance in identifying replacements. Several employers suggested a pool of newly discharged veterans. A few employers suggested providing them with a list of RC members in the area who are seeking employment so they could hire them to temporarily fill the positions while their employees were on military leave. The issue of key employees' (e.g., those in management positions) absences presenting particular challenges for employers recurred in these responses.

For longer absences, some employers would like periodic updates on the status of their employees, an ability to contact or communicate with the employees during deployment (on both personal and work-related issues), and confirmation and updates regarding the expected return date to employment:

> It would be nice if you put together a "packet" for the employer if [DoD] could provide ideas of how to keep the company and the soldier connected over the time of the employee's leave if the separation was an extended period of time. (Open-ended survey responses, question 47, CEI firm, 11–49 employees)

Several small employers also expressed a desire for resources to support the family members of activated RC members.

In open-ended survey responses, 256 employers mentioned some sort of financial benefit that would be useful for their businesses. The suggested format varied widely, depending on the nature of the employer organization. Some (presumably private) employers mentioned tax incentives, while others noted that tax incentives would not be useful for their organizations because they were government or nonprofit organizations. Several government organizations mentioned that grants to cover certain expenses would be useful. Municipal governments cited the burden that extended or frequent duty-related absences pose for local taxpayers.

Several employers highlighted the effects on business prompted by the events of 9/11. A representative from a security company participating in the open-ended surveys reported that employees of all levels and ranks had been called up since 9/11. One public FR interviewee, a large municipal employer, noted that, because it had experienced substantial duty-related absences since 2001, it explicitly planned and budgeted for overtime expenses, which are substantial. The interviewee added that the events of 9/11 inspired a great deal of "patriotism" but that, if the country is "perpetually" at war, the organization will continue to be "hit in the pocket." Indeed, multiple employers of all sizes reported experiences with deployments whose durations exceeded one year.

Several employers mentioned reimbursement for hiring, training, or retraining costs. A few employers mentioned the costs associated with unemployment insurance that stem from the need to hire temporary replacement workers and then let them go when the RC member returns. Some suggested that employers be relieved of unemployment insurance cost implications in the case of the termination of replacement workers. Others suggested that employers should be reimbursed for these costs and the cost of a severance package for the replacement worker.

Several employers asked for support in dealing with PTSD and other issues that RC members face when they return from duty. Some offered suggestions for DoD in this area:

> Our military reservist returned from his tour of duty and created significant work issues. He has finally received counseling 4 yrs. later. There needs to be much better psychological support for these folks, and continued follow-up with reservist and employers, 3–12 mos. after return. This person sued us two times, which created significant financial issues. He lost initial suit and the appeal. I don't fault the reservist. I do feel he experienced PTSD. It went untreated. Shame on the military for not being more involved with employer so we could have helped him sooner. (Open-ended survey responses, question 47, CEI firm, employer size information missing or excluded)

> In my business line of work (Public Safety) we have to always be cognizant of military reservist state of mind once they return from long deployments. It is critical to public safety and our citizens that our service members are of sound mind and body upon their return to full police duties. There has always been questions regarding our right (if we wanted to) to have returning service members seen by a psychologist to ensure that they are fit to return to police duties. This issue is lost and whose responsibility is it to delve into transitional and adjustment processes when it comes to a service member returning from a combat zone? (Open-ended survey responses, question 47, CEI firm, 500 or more employees)

> I am a retired member of the reserves. In my executive position I have ensured that our firm has gone above and beyond the requirements of USERRA. The one reservist we had returned from Iraq and was eventually terminated due to performance issues she had not display[ed] prior to mobilization. It would have been

helpful if I had a resource to which I could turn for advice regarding that dramatic change of behavior. (Open-ended survey responses, question 47, CEI firm, 100–499 employees)

Some training or an update yearly or "on request" would be most helpful. Strategically, the fire service needs some training at the chief officer level on resources available for Soldiers . . . returning with PTSD and TBI [traumatic brain injury]. The International Association of Chiefs of Police have a great transition brochure for returning soldiers and another brochure on caring for their families during deployments. You might check it out. Thanks for all that you do! (Open-ended survey responses, question 47, CEI firm, 100–499 employees)

In open-ended survey responses, 47 employers (nearly all from employers with fewer than 100 employees) specifically highlighted issues with the number and frequency of absences:

I have been the administrator at this school for 14 years. This employee has been gone on active duty for 7 of those years. That seems extreme for a reservist. That is hard on our district when we are held to such high academic standards and the person we hire is gone half the time. (Open-ended survey responses, question 19, CEI firm, 50–99 employees)

Being in the Guard now is not sustainable. Our Guardsmen are being deployed too often and too long. (Open-ended survey responses, question 19, CEI firm, 50–99 employees)

Some of these employers expressed a sentiment that the total length of military absences should be limited to prevent cases of military members being away for more time than they are at their civilian employers' sites. As previously mentioned, although USERRA protections are not available to RC members who have more than five years of cumulative military service while employed with one civilian employer, a lot of duty is exempt from that five-year limit (including some voluntary duty). Various proposals for limiting the total length of absence were provided, including limiting the total amount of duty in a particular time frame, limits to USERRA protections for voluntary duty, or requiring DoD to obtain employer input in the case of voluntary duty or duty exceeding a certain threshold in a particular time frame. Some of the employers expressing concerns about the total amount of duty also indicated that the RC member was "taking advantage" of USERRA protections to the detriment of the employer. Others highlighted challenges with total duty because of a small workforce or a high proportion of the workforce being in the RCs:

Active duty deployments should not exceed 14 months (12-month deployment with two months' training) in any five-year period. Reemployment rights are otherwise an economic and employment hardship to the employer and "temporary employee" or civilian who must be laid off. Specifically, at a previous public employer, we had a National Guard member who worked for us less than four months during a five-year period where his active duty was continually extended (he was serving stateside as a supply tech). The accumulated months of sick and annual leave as well as retirement benefits (several thousand dollars) paid with taxpayer funds, all while not working for the state and being paid for his federal service. (Open-ended survey responses, comment 2, CEI firm, 50–99 employees)

Deployments and premobilization training of one year total time with a minimum of five years between deployments. (Open-ended survey responses, comment 2, CEI firm, employer size data missing or excluded)

It would be helpful if that absence were as infrequent as possible, especially as our organization has approximately 10 percent of our workforce subject to military duty. (Open-ended survey responses, question 19, CEI firm, 100–499 employees)

Another suggestion provided by three employers in open-ended survey responses was for the federal government to provide special contracting opportunities or contracting preferences for RC employers:

We would like to have the opportunity to bid on logistics contracts for the government and have a consideration due to the fact that we support the military. (Open-ended survey responses, comment 2, CEI firm, employer size data missing or excluded)

Direct help with getting on the list for gaining govt contracts. (Open-ended survey responses, question 19, CEI firm, one to ten employees)

Afford businesses supporting National Guard or Reserve members some advantage when bidding on Federal and/or Federally funded construction projects or purchases. (Open-ended survey responses, question 19, CEI firm, 100–500 employees)

Although we have highlighted here the suggestions and complaints of employers that provided such feedback, it is also worth noting that a large number of those providing open-ended responses indicated that they sought no additional support and were prepared to do all they could to support their employees in fulfilling their military duty.

Summary

Because few U.S. employers actually employ reservists and thus experience duty-related absences, USERRA's overall impact on employers is small. Even among those employers that do experience a duty-related absence, a majority reported that duty-related absences had no impact on their business. Although there is some evidence that employers that experience longer absences were more likely than other employers to report an impact, some employers that experience duty-related absences of a year or more reported no business impact.

But, for those employers that do experience an impact, the reported magnitude can be significant. Employer sector and employer size appear to have a strong influence on employer impact, in part because these characteristics tend to constrain the methods an employer can use to deal with a duty-related absence.

Our analysis of the National Survey of Employers, employer interviews, and RC member surveys indicate that employers' perspectives on the impact of duty-related absences and measures that would help them deal with those absences vary.

Employers expressed different perspectives on the types of support that would be most useful in allowing them to deal with duty-related absences. Employers also had different opinions regarding whether absences should be shorter but more frequent or longer and less frequent. Overall, employers expressed a desire for fewer absences (total amount of time away appears to be more of an issue than how that duty is packaged), access to updated information regarding duty dates to allow for employer planning, and information to help RC members with the transition back to civilian life. The employer responses suggest that no single approach to ameliorating employer impact would improve employer support across the board.

CHAPTER FIVE

Employer Awareness of USERRA and Perspectives on ESGR

The mission of ESGR is to facilitate and promote

> a cooperative culture of employer support for National Guard and Reserve service by developing and advocating mutually beneficial initiatives; recognizing outstanding employer support; increasing awareness of applicable laws and policies; resolving potential conflicts between employers and their service members; and acting as the employers' principal advocate within DoD. (ESGR, undated [d])

In this chapter, we describe employers' awareness of their responsibilities under USERRA and perceptions of ESGR. We also describe employer suggestions for improving USERRA assistance and support for Guard and Reserve service.

Employer Contact with DoD

Although RC members are required to provide DoD with employer contact information through the CEI database, in part to support employer outreach efforts, responses to the National Survey of Employers indicate that few employers are, in fact, receiving meaningful communication from DoD. Among RC employers responding to the survey, only 17 percent reported any contact with an RC member's military supervisor in the previous 36 months. A somewhat larger percentage (25) reported any contact with ESGR in the previous 36 months.

Employer Awareness of USERRA

Responses to the National Survey of Employers indicate a lack of employer awareness about the requirements of USERRA and how ESGR can help. Although awareness is stronger among employers of RC members than among employers that do not employ RC members, that awareness is still far from complete. As reflected in Figure 5.1, 26 percent of RC employers disagreed or strongly disagreed with the statement, "I

Figure 5.1
Reserve Component Employers' Knowledge of USERRA

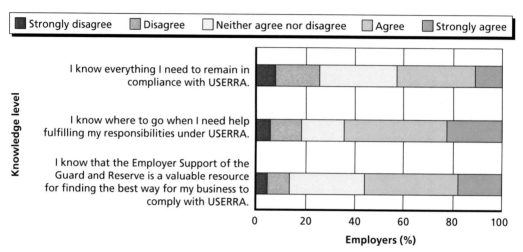

SOURCE: Authors' analysis of 2011 DoD National Survey of Employers data.
NOTE: Unweighted number of observations: Q1 = 9,486; Q2 = 9,506; and Q3 = 9,489.
RAND RR152-5.1

know everything I need to remain in compliance with USERRA." Eighteen percent disagreed or strongly disagreed with the statement, "I know where to go when I need help fulfilling my responsibilities under USERRA."

Thirteen percent of RC employers disagreed or strongly disagreed that "I know that the Employer Support for the Guard and Reserve is a valuable resource for finding the best way for my business to comply with USERRA." Thirty-one percent of RC employers responded "neither agree nor disagree" to this statement, and it is difficult to know whether that response reflects a lack of awareness of ESGR or no opinion as to whether it is a valuable resource. Open-ended survey responses reflecting experiences with ESGR were mixed, with some emphasizing that ESGR has been an extremely helpful and useful resource and others saying that ESGR is not helpful or too quick to insist that the employer has to rehire the returning RC member.

An analysis of survey responses reveals that, overall, small firms and private firms were more likely to lack the knowledge to remain in compliance with USERRA and were not aware of ESGR or where to go for information on USERRA (see Appendix B, Table B.28).

Employers that do not employ an RC member are much more likely to disagree with the above statements (see Figure 5.2). More than 45 percent do not feel that they fully understand their obligations under USERRA. Our analysis of the survey data reveals that public employers and all size categories of employers with more than ten employees are less likely to disagree with the statements than private employers and employers with one to ten employees (respectively).

Figure 5.2
Non–Reserve Component Employers' Knowledge of USERRA

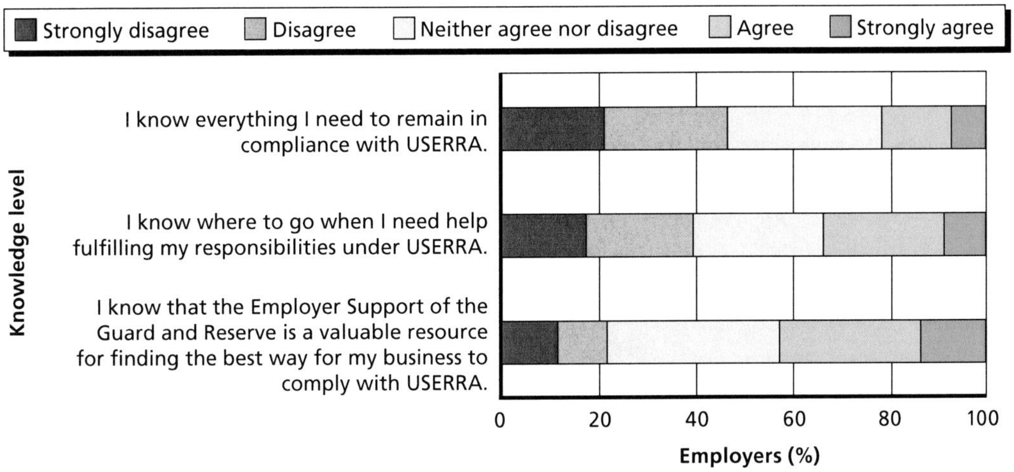

SOURCE: Authors' analysis of 2011 DoD National Survey of Employers data.
NOTE: Unweighted number of observations: Q1 = 850; Q2 = 845; and Q3 = 852.
RAND *RR152-5.2*

Of the employers in our interview pool, only six said that they were aware of and appeared to be fairly well-versed in USERRA policy and requirements; all were from the public sector. Two of the eight public employers with whose representatives we spoke expressed that they were *not* familiar with USERRA. None reported experience with disputes or recalled USERRA-based claims being filed. Those employers that were familiar with USERRA also declared themselves to be consistently in compliance and to be very supportive of returning RC members (and of military service members in general).

When asked about difficulties complying with USERRA regulations, interviewees with familiarity mentioned very few concerns or issues. One large public utility employer interviewed found it difficult to track the respective durations of its RC employees' duty-related absences but added that this was attributable to administrative processes rather than to USERRA itself. Another public-sector employer we interviewed expressed ambiguity and concern over USERRA's five-year stipulation in the face of an exorbitant number of RC deployments the municipality had experienced since 2001.

In the survey, employers were asked to rank whether certain sources of information about USERRA would be helpful or not (see Appendix B, Table B.29). Responses for all employers are described in Figure 5.3. (Responses of RC and non-RC employers were similar and are not broken out separately.) The sources rated as most helpful were a fact sheet on the web or a call center to answer questions. Among the sources rated as least helpful were a workshop with an ESGR trainer and other business owners, a visit

Figure 5.3
Usefulness of Potential Sources of Information on USERRA, All Employers

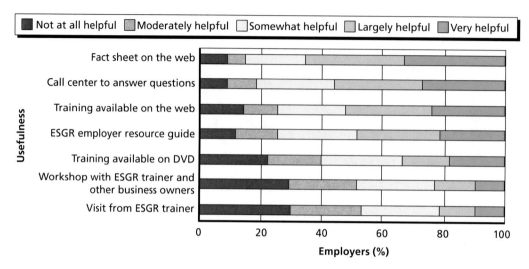

SOURCE: Authors' analysis of 2011 DoD National Survey of Employers data.
RAND *RR152-5.3*

from an ESGR trainer, and training available on a DVD. Web training was ranked as somewhat more helpful. Forty-five percent of employers would have liked information about USERRA upon request, while 17 percent preferred to receive information when an employee must be absent for military duties. Twenty-nine percent reported that their businesses had no need for information or training about USERRA, and 9 percent reported that they would like yearly information or training. A few survey respondents emphasized the overall finding that it would be most helpful to have information available on an as-needed basis. Smaller employers in particular reported not wanting to be bothered with information about the law until they are faced with a duty-related absence.

Employer Awareness of ESGR Programs

Overall, awareness of ESGR programs is very limited, and interestingly, RC employers were not substantially more likely than non-RC employers to be aware of programs. We begin with a summary of findings from the analysis of National Survey of Employers data.

Statement of Support

Examining this factor by employer status as an RC employer (results not shown), we found that 65 percent of firms that employed RC members were not aware of the SOS for RC members provided by ESGR; 91 percent of non-RC employers were not aware. Twenty-one percent of RC employers reported displaying a signed SOS.

Programs and Contact with ESGR

All employers were asked about their awareness of and participation in ESGR programs: unit mission ceremony, military installation visit, boss lift, Yellow Ribbon program, lunch or breakfast with the boss, or "other." Overall, awareness and participation were low, even among RC employers. As reflected in Table 5.1, fewer than 5 percent of RC employers reported participating in any particular program, and fewer than 17 percent confirmed awareness of any of the programs. Among non-RC employers, fewer than 10 percent were aware of any of the ESGR programs.

Among RC employers, 17 percent reported that their businesses had some contact with their RC employees' military supervisors in the past 36 months. Fifty-eight percent reported no contact, and 25 percent were unsure. Sixteen percent reported that they had had contact with ESGR in the past 36 months by letter or brochure; 10 percent had visited the ESGR website. Most RC employers, however, reported no contact with ESGR through these means or did not know whether contact had occurred.

Awards

RC employers were asked whether their businesses had received any awards in the past 36 months for the support provided to RC employees, such as the Freedom Award, Above and Beyond Award, or Patriot Award. Less than 5 percent had received any of the awards, with the exception of the Patriot Award, which 5 percent reported receiving. At least one-quarter of employers were unsure whether their businesses had received the award for all awards listed. (See Table B.32 in Appendix B.)

Employers seemed interested in learning more about these programs, with some survey respondents writing in the "other" comments that they wanted additional information:

> Did not know ESGR existed. Shouldn't it be mentioned as a resource on the employer's copy of the military orders?

Only two of the 16 employers with whose representatives we spoke were aware of ESGR and the programs it offers. Both of these employers—one of which was in the public sector, and the other of which was a small firm—were in fields related to law enforcement and security protection. The public-sector interviewee with awareness of these programs was also a recipient of various awards in the past. Also, one employer purported to be unfamiliar with the programs *but* said that the firm had previously received an award or recognition.

Among our interviewees, few had heard of ESGR, but some were curious to learn more about the programs. One public employer surmised that, despite having more

Table 5.1
Employers' Awareness of and Participation in ESGR Programs (%)

Awareness Level	Unit Mission Ceremony	Military Installation Visit	Boss Lift	Yellow Ribbon Program Activities	Lunch or Breakfast with the Boss	Other
RC						
Yes, and my business has participated in this program in the past 36 months	4	4	3	4	2	1
Yes, but my business has not participated in this program in the past 36 months	10	11	9	12	8	5
No, my business was unaware of this program	47	47	49	47	50	44
I do not know whether my business is aware of this program	39	38	39	38	39	50
Non-RC						
Yes, and my business has participated in this program in the past 36 months	1	1	1	2	1	1
Yes, but my business has not participated in this program in the past 36 months	7	6	5	7	6	4
No, my business was unaware of this program	55	57	56	54	56	50
I do not know whether my business is aware of this program	37	36	38	37	38	45

SOURCE: Authors' analysis of 2011 DoD National Survey of Employers data.

NOTE: Margin of error ranges from ±0.4 percent to ±6.5 percent. Because of rounding, some percentages do not sum to 100. RC weighted Ns: ceremony = 164,759; visit = 164,453; lift = 163,728; Yellow Ribbon = 164,129; lunch = 164,453; other = 85,276. RC unweighted Ns: ceremony = 9,407; visit = 9,388; lift = 9,359; Yellow Ribbon = 9,376; lunch = 9,401; other = 4,844. Non-RC weighted Ns: ceremony = 6,833,989; visit = 6,837,071; lift = 6,826,215; Yellow Ribbon = 6,833,989; lunch = 6,824,283; other = 3,508,657. Non-RC unweighted Ns: ceremony = 845; visit = 846; lift = 844; Yellow Ribbon = 845; lunch = 843; other = 413.

than 500 employees, the programs might not be beneficial to it, based on the relatively small number of RC employees it had at the time. This and other interviewees from the public sector expressed interest in learning more about the various recognition programs. Several employers suggested that more communication in general between

ESGR and employers would be positive for business or relations with RC members. More specifically, a large public employer stated that simply knowing *whom* to talk to would be a step in the right direction for the municipality.

Reserve Component Member Awareness and Perspectives on ESGR

The 2011 SOFS-R asked RC members about their perspectives on ESGR. As with employers, few RC members reported having contact with ESGR. Of all RC members surveyed in January 2011, 12 percent reported contacting ESGR at some point; 59 percent of those who had problems with civilian employment upon returning from their most recent activation reported contacting ESGR for help. Of those who reported contacting ESGR for help, 36 percent reported being very satisfied with the promptness of ESGR's response to their USERRA problems; 30 percent reported being very unsatisfied or unsatisfied. Fifty-three percent reported being satisfied or very satisfied with the manner in which their requests for assistance were handled; 30 percent reported being very unsatisfied or unsatisfied.

CHAPTER SIX

Conclusions and Recommendations

USERRA provides two basic types of rights to service members: protection from employment discrimination based on military service, and an entitlement to reemployment after an absence due to military service, provided that certain conditions articulated in the law are met. The law reflects a strong congressional commitment to support employment for service members, recognizing full well that such protections could impose costs on employers. In according these protections, the law explicitly covers not just involuntary but also voluntary duty. In passing USERRA, Congress considered whether to exempt small employers from the legislation and decided not to. However, the law does include exceptions to the reemployment requirements in cases in which they would impose undue hardship on the employer or in which business conditions had changed so that it would be unreasonable or impossible to honor those requirements.

DoD has long recognized that the existence of USERRA itself is not sufficient to guarantee employer support for RC members and their military duty. As with any labor law, compliance and enforcement can only go so far in ensuring that RC members receive the benefits they are due under the law. Employer support for the law's provisions is key. Support for USERRA is developed though DoD's RC utilization policies that seek to ensure the judicious use of RC members and through ESGR and RC support programs and outreach efforts that work to recognize and support employers.

Shifts in DoD policy regarding the utilization of RC members since 1994 have led to an overall increase in the number of duty days for RC members and may have led to more-frequent utilization of particular categories of RC members.

Our analysis suggests that, in spite of the increases in RC utilization, duty-related absences do not pose serious issues for most employers. First of all, a very small percentage of U.S. employers employ RC members at any given point in time, and most of those that do employ RC members employ few of them. Although individual RC members are quite likely to experience duty-related absences in the current environment, from the perspective of an employer, a duty-related absence is a rare occurrence. In contrast, almost all employees working for employers covered by the FMLA are potentially eligible to take an FMLA leave.

Even among employers that have experienced duty-related leaves, only one-third reported any change in their business operations as a result of those absences. At the same time, our analysis suggests that duty-related absences can pose a substantial burden on employers in some circumstances. Our study provides some insights into the factors that influence that effect and what measures could help employers deal with those effects.

As described in Chapter One, our study focused on the following research questions:

- What are the legal protections provided by USERRA, what obligations do they impose on employers, and what are the areas of ambiguity?
- To what extent do employers understand their obligations under USERRA?
- What factors influence the impact that USERRA protections can have on employers?
- What changes to USERRA or to DoD policy would employers find useful in fulfilling their obligations under USERRA?

Legal Protections Provided by USERRA Are Generally Clear and Consistent with Other Employment Laws

We considered whether there is sufficient need for a rewrite of the USERRA legislation based on USERRA's legal and legislative history. USERRA is generally clear and consistent with other employment laws. Therefore, we conclude that there is no need for substantial revision to the legislation. In this section, we highlight some areas of difficulty or ambiguity in the law that may require some action. These issues are relatively minor and do not get at the core antidiscrimination and reemployment protections provided by USERRA. The issues we mention are likely to be resolved in due course as USERRA cases wind their way through the courts, but Congress may see fit to adjust the current law in order to clarify these issues more quickly and directly.

USERRA has benefited from a long history of service member rights legislation and legislative revisions over many years. Although there are some legal challenges that arise from a new reliance on waivers and an adoption of alternative dispute resolution options, the legal issues surrounding USERRA's application do not appear to justify a rewrite of the law. However, the change in military policy toward greater reliance on the RCs may stress civilian employers beyond the levels contemplated by Congress. If so, Congress ought to consider whether the allocation of responsibilities and burdens under USERRA reflects congressional intent and serves public policy.

Understanding of USERRA, Employer Obligations, and Where to Go for Help Is Incomplete

Our analysis reveals that employer knowledge of USERRA and ESGR is incomplete. Although RC employers express greater familiarity with USERRA, confidence in their ability to comply with the law, and awareness of ESGR, roughly one-quarter of RC employers disagreed with the statement that they knew all they needed to know in order to remain in compliance with USERRA. Small employers were more likely to disagree with this statement than larger employers. Well below half of RC employers were aware of the SOS and of programs and awards sponsored by ESGR. Many employers expressed an interest in the types of programs, awards, and supports offered by ESGR.

The Impact of USERRA Protections on Employers Is Highly Varied and Influenced by a Wide Range of Factors

The impact of USERRA protections stems from the direct and indirect implications for business operations due to the duty-related absence of RC members from their employment. These may include the need to manage workload, the need to replace the RC member, loss of business, costs of health and other benefits, and lost productivity. Among employers that experience a business impact, issues related to managing workload are most common. However, at least some employers reported each type of impact, and small employers were more likely than other employers to report a loss of business.

Our study confirms that factors suggested by prior research and described in our conceptual framework are relevant to employer impact. There is evidence that longer or more-frequent absences increase the impact on employers. There is evidence that the absences of highly skilled employees, key personnel, or employees who are "one of a kind" in their organizations have a greater impact. There is evidence that smaller employers are more likely to experience impacts and that other employer characteristics play a role as well. However, our analysis does not identify any single factor as a key driver of impact. For example, although employers that experience longer absences are more likely than other employers to report an impact, many employers that experienced absences longer than a year reported no impact.

Instead, our analysis suggests that it is a subtler combination of factors, including the employer circumstances at the time of activation, that results in a significant employer impact. Special circumstances faced by school districts, by businesses depending on personal relationships with clients or customers, and by businesses that rely on highly skilled or trained employees can cause particular problems or lead to a need for more advance notice to plan for absences. In addition, employers reported challenges in dealing with the activation of the "only" employee of a certain type in the organization—whether or not that individual is highly skilled. Because it is not possible to shift the work to another employee in such situations, more advance planning is needed in these situations.

Employers report a need to plan for the return of RC members, as well as for their pending absences. Some employers indicated that planning for the return is inhibited by a lack of information on whether RC members will return to work after duty and the duty end date. For highly skilled employees, an employer can face a need to retrain an employee when he or she returns, particularly in the case of long absences. Some employers mentioned challenges involved in terminating replacement workers, including increased costs of unemployment insurance for the organization as a result. In addition, about 15 percent of employers that experienced duty-related absences reported that returning RC members faced problems, such as difficulty interacting with clients or emotional problems, upon return.

No Single Change to DoD Policy or Support Programs Would Address the Concerns of All Employers

The conceptual framework we presented in Chapter Three posited that DoD policy or support programs could ameliorate employer impact by influencing the nature of absences or possibly the types of RC members who are activated. Notably, enforcing DoD policy limiting involuntary mobilization to a ratio of one year mobilized to five years demobilized (R. Gates, 2007) or linking preactivation training directly with any activation of one year or more (a contiguous-training approach) would make it easier for employers to deal with a duty-related absence. Our analysis revealed that no one structure of military duty is preferred by all employers. Most employers do not state a preference for or against shorter, more-frequent duty (relative to longer, less-frequent duty). And although two-thirds of all employers expressed a preference for a contiguous-training approach in the event of a long activation, one-third of employers preferred to have a gap between training and deployment.

Similarly, there was no consensus among employers as to the support programs that would be most useful to their businesses in the event of duty-related absences. The largest share of employers reported that no support programs were needed, but each of the survey options was selected by at least one employer. Employers did express a desire for advance notification of both absence and return, and many employers wanted clear and consistent documentation of the term of duty directly from DoD. Some employers—most notably, school systems, employers with seasonal or cyclical work schedules, smaller employers, and employers facing the absences of key personnel or multiple personnel—requested more flexibility and employer input in scheduling both training duty and deployment. Other common suggestions provided by employers were limits to voluntary duty, replacement assistance (possibly at government expense), government contracting preferences for RC employers, and a way to stay in contact with RC members during their terms of duty.

Recommendations

Providing Employers with More Flexibility Holds More Promise Than Broad Changes to USERRA or Utilization Policy

Our findings suggest that across-the-board changes to USERRA, utilization policy, or duty structure applied to all RC members and all employers will not resolve employer impacts uniformly, given the sporadic and varied nature of problems faced by employers.

The analysis reveals that employer impact varies widely in ways that cannot be easily anticipated or predicted based on employer characteristics. Yet, at times, that impact can be significant. For this reason, we recommend that DoD consider implementing a more extensive and systematic appeal process that would not be limited to specific categories of employers but rather could be accessed by employers on an as-needed basis. Such an appeal process exists in the United Kingdom. Adopting a systematic appeal process would allow DoD to effectively address the most-significant instances of employer impact and balance the needs of DoD with those of the employer. DoD already allows individual RC members to petition for relief from activation. Providing employers with access to a similar process could ameliorate the most-significant instances of employer impact.

In fact, such flexibility is already provided on a very limited basis. Under the "key personnel" provision of USERRA, employers are allowed to petition to have employees who hold key positions moved from the active reserve to the inactive reserve. In addition, DoD considers requests from employers for the deferral of duty. Each type of request is rare, but, based on the information we obtained, they are typically granted when received.

We conclude that a more explicit appeal process, although it would impose some administrative burden on DoD, would have some advantages. USERRA already includes provisions that exempt an employer from reemployment requirements in the event that they would pose undue hardship on the employer. However, utilizing these exemptions after the duty has occurred has undesirable features for both the RC member and the employer. For the employer, it raises the possibility of a legal challenge from the RC member, who might disagree with the undue-hardship claim. For the RC member, this approach puts him or her in a situation in which he or she may leave for duty assuming that he or she will have a job upon return, only to find that this is not the case. An up-front appeal process that allows employers to make the case in advance that hardship is likely to occur would reduce uncertainty for both sides. If DoD defers or exempts the RC member from duty, the employer would avoid the absence and the resulting hardship. The RC member could presumably still choose to undertake the duty but with the full understanding that he or she is waiving USERRA reemployment protections. Another advantage of such a process is that it would allow DoD to track information on the reasons for hardship and the types of employers that experience hardship. That information could then be used by DoD to develop targeted support programs.

DoD is currently in the process of reviewing and reissuing its policy on employers' ability to petition the RCs for adjustments to their employees' duty-related absences. DoDI 1205.12 established a policy permitting adjustments when a duty-related absence would affect a civilian employer's business. But, at the time of our interviews, there was little knowledge of the policy among the employers we contacted. Moreover, it is unclear how well established the policy is at the service level. Greater clarity in terms of who is responsible for processing employer-requested adjustments, standards for reviewing and either granting or denying such adjustments, and adequate record-keeping would improve the implementation of the policy. Furthermore, such revisions would move DoD policy in the direction of increased flexibility to account for adverse impacts on employers' businesses. Because most of the survey respondents did not experience adverse effects from duty-related absences, the effect that a more explicit policy might have on DoD may be small, while the avoidance of undue economic hardships felt by adversely affected businesses may be significant.

DoD Should Explore Ways to Standardize and Expand Communication with Employers About Duty-Related Absences and Available Resources

Many employers would welcome improvements to communication from DoD about duty-related absences. Although many employers rely on RC members to update them about absences, some employers require documentation from DoD for benefit processing purposes or simply to validate or update the information that has been provided to them by RC members. Although USERRA requires an employee to notify his or her employer about military duty, the documentation requirements are flexible and are readily waived. As a practical matter, it would be difficult for an employer to deny reemployment because of a lack of documentation of the military duty, given the provisions of USERRA.

Our findings suggest that ESGR should continue its efforts to promote awareness of USERRA and provide support for employers in addressing their USERRA obligations. In view of the fact that a small minority of employers experience duty-related absences at any given point in time, ESGR should consider targeting communication with employers toward those who are experiencing duty-related absences. The EN system in the United Kingdom provides an example of how communication could be standardized and targeted.

DoD could address employer concerns by developing and preparing a simple, easy-to-read form for employers that documents duty start and end dates. The form could include information on the time frame in which the RC member must notify the employer of his or her intention to return (based on the length of duty) per USERRA provisions. Importantly, DoD should provide employers with updated information any time duty has been extended for the benefit of those employers that are not receiving updates directly from the RC members. As suggested above, DoD could also use these notifications as an opportunity to provide a USERRA information packet and ESGR contact information to employers at the start of duty and information on DoD resources for service members and veterans at the end of duty.

To support some of these outreach efforts, DoD would need to reconsider existing limitations on the use of CEI data in ways that identify RC members. If those limitations were relaxed, DoD could institute a process through which RC members could opt out of providing such information, like in the UK EN system. In addition, further improvements to the CEI database and RC member submissions and updates to it may be needed. In its 2007 report, GAO raised concerns about the quality of information in the CEI database. DoD updated some of its procedures in response to that GAO report. However, responses to the National Survey of Employers suggest that limitations persist.

ESGR may also want to consider partnering with DOL and the U.S. Small Business Administration (SBA) to improve the dissemination of information about USERRA to smaller employers. USERRA is a labor law that protects service members against employment discrimination and provides them with reemployment rights in the event of duty-related leave. From an employer perspective, USERRA is an anti-discrimination employment law and a law governing employee leave practices. Rather than describing USERRA as a law related to the military, ESGR should consider emphasizing the antidiscrimination and leave protection aspects of the law. Unlike other employment laws governing employee leave (notably, the FMLA), USERRA applies to all employers regardless of size. ESGR should recognize that employers with fewer than 50 employees might be less likely to be aware of the reemployment obligations they face under USERRA.

Because employers with fewer than 50 employees are not required to provide leave under the FMLA, these employers are less likely to have experience dealing with extended unpaid employee leaves. ESGR should consider working with SBA to develop a network for mentors and a description of best practices that smaller employers have used to effectively deal with duty-related leave.

As part of these efforts, DoD should make sure to enhance employer awareness of programs and resources available to help employers and RC members deal with common issues that arise in the transition back to civilian life after a deployment (Werber et al., 2008). To the extent that DoD has developed resources for active-duty military members who are deployed, ensuring that these resources are available to and possibly modified to suit the needs of RC members is critical. Most RC employers value the RC members they employ and are supportive of their service to the United

States. In cases in which an RC member returns from a deployment with problems, employers want to help. DoD should improve access to information on support programs that might be available to RC members. DoD should also explore why federal employers are more likely than other employers to report that employees are returning from duty with employment-related challenges. It may be that federal employers have a higher level of awareness of possible issues or that RC members who are federal employees are experiencing such challenges at a higher rate.

Of course, these suggestions will be useful to employers only to the extent that DoD maintains up-to-date information about employers of RC members. Responses to the National Survey of Employers suggest that continued efforts to improve this information are needed.

On a related note, DoD should consider options for allowing employers to send messages to service members who are on duty status and, with RC member agreement, engage in direct communication with them. Other options, such as an ability to subscribe to periodic updates on unit activities, would be of interest to some employers. Expanding the options for employers to remain in connection with their RC member employees, either directly or indirectly, during a period of duty may be particularly desirable for smaller employers.

Employer Peer Resource Networks Could Assist Employers in Dealing with Duty-Related Absences

Employers that have had successful experiences with duty-related absences can be a useful resource for other companies facing such absences. A network of employers of different sizes and in different industries and sectors could be a powerful peer resource. DoD should consider soliciting employers to serve as peer mentors and developing approaches similar to existing award programs for recognizing these employers. Such a program would provide another avenue for recognizing employers and building involvement while providing direct support.

As described in this report, employers offered other suggestions for support that appear reasonable but might be valued by a relatively small number of employers. ESGR should explore the feasibility of providing such supports and further assess employer perspectives on various alternatives. These include direct assistance for identifying replacement workers (possibly at no or reduced cost to the employer), government contracting preferences for RC employers, and adjustments to the way in which the separation of replacement workers for RC members is accounted for in calculating unemployment insurance costs imposed on employers.

Glossary

activation. Order to active duty for purposes other than training. May be voluntary or involuntary. May or may not be associated with a deployment.

Civilian Employment Information. A program requiring Reserve Component members to register information about their civilian employers and job skills with the Department of Defense for inclusion in a department-wide database. The employer information is used to give consideration to civilian employment necessary to maintain national health, safety, and interest when considering members for recall; ensure that members with critical civilian skills are not retained in numbers beyond those needed for those skills; and inform employers of reservists of their rights and responsibilities under the Uniformed Services Employment and Reemployment Rights Act.

deployment. Movement of forces for the purposes of a military operation.

Employer Support of the Guard and Reserve. A Department of Defense field operating agency established in 1972 to help Reserve Component service members and their civilian employers.

escalator rule. A court-created legal doctrine that entitles a service member returning to civilian employment after military duty to resume employment at the level at which he or she would have been employed had the employment not been interrupted by military service, with some exceptions. This includes most seniority-related benefits.

first responder. Personnel working in emergency medical services, explosives, firefighting, handling incidents involving hazardous materials, law enforcement, and search and rescue.

full time. In the Status of Forces Surveys, this is defined as 35 or more hours per week.

National Guard. Armed forces governed by each state's governor for such efforts as assistance in natural disasters, state emergencies, and civil unrest. The President can call the forces to active service, in which case they report to the President.

National Survey of Employers. Designed to determine ways of supporting civilian employers when their National Guard and Reserve Component employees are absent because of military duties, determine general attitudes toward those employees and their contributions to employers, and examine knowledge of and compliance with the Uniformed Services Employment and Reemployment Rights Act. Groupings of employers by size, data available for employers in the Civilian Employment Information database, are one to ten, 11–49, 50–99, 100–499, and 500 or more employees at all U.S. locations. The survey also distinguishes between employers that are first-responder organizations and those that are not and among public (i.e., government), nonprofit, and for-profit employers.

Reserve Component. The seven reserve components of the U.S. armed forces, consisting of the U.S. Army Reserve, U.S. Navy Reserve, U.S. Air Force Reserve, U.S. Marine Corps Reserve, Army National Guard, Air National Guard, Air Force Reserve, and Coast Guard Reserve. These organizations are often collectively referred to as *the Guard and Reserve*; in this report, we use the term *Reserve Components*.

Reserve Component employer. A civilian employer that has employees who serve as part of a Reserve Component. A non–Reserve Component employer is one that has no employees in a Reserve Component, as far as it is aware.

Status of Forces Survey. Conducted by the Defense Manpower Data Center for the Reserve Components, Active Components, and civilian employees to help the Department of Defense evaluate existing programs and policies, establish baselines before implementing new programs and policies, and monitor the progress of programs and policies and their effects on the total force. This survey categorizes the largest employers as those with 500 or more employees at all U.S. locations; the next category down is employers with 100–499 employees at all U.S. locations. In this survey, small businesses are those with ten to 100 employees at all U.S. locations, and very small businesses are those with one to nine employees at all U.S. locations. The survey also distinguishes among private, nonprofit, federal, state, local, self-employed, and employed and paid by a family business. In our analyses of these data, we treat those in the latter two categories as not working for an employer.

Bibliography

"Absence Survey Finds Costs Are Up, Rate Steady," *Benefits and Compensation Digest*, Vol. 43, No. 1, January 2006, p. 18.

Banks, Hannah, "*Staub v. Proctor Hospital*: Cleaning Up the Cat's Paw," *Duke Journal of Constitutional Law and Public Policy Sidebar*, Vol. 6, 2011, pp. 71–92. As of March 19, 2013: http://scholarship.law.duke.edu/cgi/viewcontent.cgi?article=1067&context=djclpp_sidebar

Bates, Steve, "2004 Study: FMLA Cost Employers $21B," *HR Magazine,* Vol. 50, No. 6, June 2005, pp. 36–42.

Bettenhausen, Laura, "The FAA and the USERRA: Pro-Arbitration Policies Can Undermine Federal Protection of Military Personnel," *Journal of Dispute Resolution*, Vol. 2007, No. 1, 2007, pp. 267–282.

Buck, Jennifer C., Thomas L. Bush, Barbara A. Bicksler, Karen I. McKenney, and John D. Winkler, "The New Guard and Reserve," in John D. Winkler and Barbara A. Bicksler, eds., *The New Guard and Reserve*, San Ramon, Calif.: Falcon Books, 2008, pp. 3–14. As of March 19, 2013: http://www.sainc.com/reports/pdf/New_Guard_Reserve.pdf

Bugbee, Daniel J., "Employers Beware: Violating USERRA Through Improper Pre-Employment Inquiries," *Chapman Law Review*, Vol. 12, No. 2, Fall 2008, pp. 279–300.

Busby, Colin, *Supporting Employees Who Deploy: The Case for Financial Assistance to Employer of Military Reservists*, Toronto, Ont.: C. D. Howe Institute, Backgrounder 123, January 2010. As of March 19, 2013: http://www.cdhowe.org/pdf/backgrounder_123.pdf

Chu, David S. C., "Civilian Employment Information (CEI) Program," memorandum for secretaries of the military departments; chair of the Joint Chiefs of Staff; Under Secretaries of Defense; comptroller of the U.S. Department of Defense; director, Defense Research and Engineering; assistant secretaries of defense; general counsel of the U.S. Department of Defense; inspector general of the U.S. Department of Defense; director, Operational Test and Evaluation; assistants to the Secretary of Defense; director of administration and management; and directors of defense agencies, Washington, D.C.: Under Secretary of Defense for Personnel and Readiness, March 21, 2003. As of March 19, 2013: http://ra.defense.gov/documents/main/CEIChu.pdf

Code of Federal Regulations, Title 32, National Defense, Part 44, Screening the Ready Reserve. As of March 25, 2013: http://www.ecfr.gov/cgi-bin/retrieveECFR?gp=&SID=664ce0e85345d811809f23465362bfc5&r=PART&n=32y1.1.1.4.23

Crotty, Matt, "The Uniformed Services Employment and Reemployment Rights Act and Washington State's Veteran's Affairs Statute: Still Short on Protecting Reservists from Hiring Discrimination," *Gonzaga Law Review*, Vol. 43, No. 1, August 2007, pp. 169–197. As of March 19, 2013:
http://blogs.gonzaga.edu/gulawreview/files/2011/02/Crotty.pdf

Darby, Larry F., and Joseph P. Fuhr, *Benefits and Costs of the Family and Medical Leave Act of 1993: A Consumer Welfare Perspective*, Darby Associates, February 16, 2007. As of August 31, 2010:
http://www.protectfamilyleave.org/research/darby_fmla.pdf

Defence Reserves Support, "About ESPS Payments," undated (a), referenced August 20, 2012. As of March 26, 2013:
http://www.defencereservessupport.gov.au/for-employers/
esps-payments-to-support-employers-of-reservists/about-esps-payments.aspx

———, "Substantial Financial Hardship or Loss Claims," undated (b), referenced August 1, 2012. As of March 26, 2013:
http://www.defencereservessupport.gov.au/for-employers/
esps-payments-to-support-employers-of-reservists/substantial-financial-hardship-or-loss-claims.aspx

Defense Manpower Data Center, *June 2008 Status of Forces Survey of Reserve Component Members: Tabulations of Responses*, Arlington, Va., Report 2009-008, 2009.

DePremio, Heather, "The War Within the War: Notice Issues for Veteran Reemployment," *Naval Law Review*, Vol. 53, 2006, pp. 31–54.

Dixon, Lloyd, Susan M. Gates, Kanika Kapur, Seth A. Seabury, and Eric Talley, "The Impact of Regulation and Litigation on Small Businesses and Entrepreneurship: An Overview," in Susan M. Gates and Kristin J. Leuschner, eds., *In the Name of Entrepreneurship? The Logic and Effects of Special Regulatory Treatment for Small Business*, Santa Monica, Calif.: RAND Corporation, MG-663-EMKF, 2007, pp. 17–68. As of March 20, 2013:
http://www.rand.org/pubs/monographs/MG663.html

DMDC—*See* Defense Manpower Data Center.

DoD—*See* U.S. Department of Defense.

DoDI 1205.12—*See* Under Secretary of Defense for Personnel and Readiness, 1997.

DOL—*See* U.S. Department of Labor.

Doyle, Colin M., and Neil Singer, "The Effect of Reserve Component Mobilizations on Civilian Employers," in John D. Winkler and Barbara A. Bicksler, eds., *The New Guard and Reserve*, San Ramon, Calif.: Falcon Books, 2008, pp. 135–152.

Employer Partnership of the Armed Forces, "Overview," undated. Referenced March 26, 2013.

Employer Support of the Guard and Reserve, "Above and Beyond Award," undated (a). As of March 26, 2013:
http://www.esgr.mil/Employer-Awards/Above-and-Beyond-Award.aspx

———, "Employer Awards," undated (b). As of April 8, 2013:
http://www.esgr.mil/Employer-Awards/ESGR-Awards-Programs.aspx

———, "Extraordinary Employer Support Award," undated (c). As of March 26, 2013:
http://www.esgr.mil/Employer-Awards/Extraordinary-Support.aspx

———, "Mission and Strategy," undated (d), referenced November 2, 2012. As of March 26, 2013:
http://www.esgr.mil/About-ESGR/Mission-and-Strategy.aspx

————, "Pro Patria Award," undated (e). As of March 26, 2013:
http://www.esgr.mil/Employer-Awards/Pro-Patria-Award.aspx

————, "Statement of Support Program, undated (f). As of April 4, 2013:
http://www.esgr.mil/Employers/Statement-of-Support.aspx

EP—*See* Employer Partnership of the Armed Forces.

ESGR—*See* Employer Support of the Guard and Reserve.

Falk, Justin Robert, *Comparing the Compensation of Federal and Private-Sector Employees*, Washington, D.C.: Congressional Budget Office, January 2012. As of March 20, 2013:
http://purl.fdlp.gov/GPO/gpo18432

Fishgold v. Sullivan Dry Dock & Repair Corporation, 328 U.S. 275, 1946.

Forte, Michele A., "Reemployment Rights for the Guard and Reserve: Will Civilian Employers Pay the Price for National Defense?" *Air Force Law Review*, Vol. 59, No. 1, 2007, pp. 287–344.

Foster v. Dravo Corp., 420 U.S. 92, 1975.

Ganapati, Priya, "FMLA Costs Hit $21 Billion in 2004," *Inc.*, April 28, 2005. As of August 30, 2012:
http://www.inc.com/news/articles/200504/fmlastudy.html

Gates, Robert, "Utilization of the Total Force," memorandum for secretaries of the military departments, chair of the Joint Chiefs of Staff, and Under Secretaries of Defense, Washington, D.C.: Secretary of Defense, January 19, 2007. As of March 20, 2013:
http://www.armyg1.army.mil/MilitaryPersonnel/Hyperlinks/Adobe%20Files/
OSD%20Memo%20dtd%2020070119%20-%20Utilization%20of%20the%20Force.pdf

Gates, Susan M., Geoffrey McGovern, Ivan Waggoner, John D. Winkler, Ashley Pierson, Lauren Andrews, and Peter Buryk, *Supporting Employers in the Reserve Operational Forces Era: Appendixes*, Santa Monica, Calif.: RAND Corporation, RR-152/1-OSD, 2013. As of summer 2013:
http://www.rand.org/pubs/research_reports/RR152z1.html

Gaudine, Alice P., and Alan M. Saks, "Effects of an Absenteeism Feedback Intervention on Employee Absence Behavior," *Journal of Organizational Behavior,* Vol. 22, No. 1, February 2001, pp. 15–29.

Golding, Heidi L. W., *The Effects of Reserve Call-Ups on Civilian Employers*, Washington, D.C.: Congressional Budget Office, May 2005. As of March 20, 2013:
http://purl.access.gpo.gov/GPO/LPS72652

Hansen, Michael L., Celeste Gventer, John D. Winkler, and Kristy N. Kamarck, *Reshaping the Army's Active and Reserve Components*, Santa Monica, Calif.: RAND Corporation, MG-961-OSD, 2011. As of March 20, 2013:
http://www.rand.org/pubs/monographs/MG961.html

Hardy, Sean M., "A Fighting Chance: The Proposed Servicemembers Access to Justice Act and Its Potential Effects on Binding Arbitration Agreements," *Pepperdine Dispute Resolution Law Journal*, Vol. 10, No. 2, 2010, Article 5. As of March 20, 2013:
http://digitalcommons.pepperdine.edu/cgi/viewcontent.cgi?article=1040&context=drlj

Harrell, Margaret C., and Nancy Berglass, *Employing America's Veterans: Perspectives from Businesses*, Washington, D.C.: Center for a New American Security, June 11, 2012. As of March 20, 2013:
http://www.cnas.org/employingamericasveterans

Hope, John B., Douglas B. Christman, and Patrick C. Mackin, "An Analysis of the Effect of Reserve Activation on Small Business," Washington, D.C.: Small Business Administration Office of Advocacy, October 2009. As of March 20, 2013:
http://archive.sba.gov/advo/research/rs352tot.pdf

Kepner, Emily M., "True to the Fable? Examining the Appropriate Reach of Cat's Paw Liability," *Seventh Circuit Review*, Vol. 5, No. 1, Fall 2009, pp. 108–158. As of March 20, 2013:
http://www.kentlaw.iit.edu/Documents/Academic%20Programs/7CR/v5-1/kepner.pdf

Lee, Konrad S., Patrick W. Fitzgerald, Daniel Peterson, and Matthew I. Thue, "Emerging Limitations of the Uniformed Services Employment and Reemployment Act," *Loyola Law Review*, Vol. 55, Spring 2009, pp. 23–44.

Leonard, Bill, "SHRM Survey Highlights Problems with FMLA," *HR Magazine,* Vol. 52, No. 8, August 2007, p. 28.

Manson, H. Craig, "The Uniformed Services Employment and Reemployment Rights Act of 1994," *Air Force Law Review*, Vol. 47, 1999, pp. 55–88.

Navarro, Chris, and Cara Bass, "The Cost of Employee Absenteeism," *Compensation and Benefits Review*, Vol. 38, No. 6, December 2006, pp. 26–30.

Nieva, Veronica, Wayne Hintze, and John Rauch, *1999 Employer Reservist Survey: Final Report,* Rockville, Md.: Westat, August 2000.

Obama, Barack, "Ensuring the Uniformed Services Employment and Reemployment Rights Act (USERRA) Protections," memorandum for the heads of executive departments and agencies, July 19, 2012. As of November 1, 2012:
http://www.whitehouse.gov/the-press-office/2012/07/19/
presidential-memorandum-uniformed-services-employment-and-reemployment-r

Orme, Geoffrey J., and James Kehoe, "Perceptions of Deployment of Australian Army Reservists by Their Employers," *Military Medicine*, Vol. 177, No. 8, August 2012, pp. 894–900.

Pauly, Mark. V., Sean Nicholson, Judy Xu, Dan Polsky, Patricia M. Danzon, James F. Murray, and Marc L. Berger, "A General Model of the Impact of Absenteeism on Employers and Employees," *Health Economics*, Vol. 11, No. 3, 2002, pp. 221–231.

Phillips, Bruce D., "The Economic Costs of Expanding the Family and Medical Leave Act to Small Business," *Business Economics*, Vol. 37, No. 2, April 2002, p. 44.

Pint, Ellen M., Amy Richardson, Bryan W. Hallmark, Scott Epstein, and Albert L. Benson, *Employer Partnership Program Analysis of Alternatives*, Santa Monica, Calif.: RAND Corporation, TR-1005-A, 2012. As of March 20, 2013:
http://www.rand.org/pubs/technical_reports/TR1005.html

Public Law 75-718, Fair Labor Standards Act, June 25, 1938.

Public Law 76-783, Selective Training and Service Act, September 16, 1940.

Public Law 88-352, Civil Rights Act, July 2, 1964.

Public Law 90-202, Age Discrimination in Employment Act, December 15, 1967.

Public Law 92-540, Vietnam Era Veterans' Readjustment Assistance Act, October 24, 1972.

Public Law 95-555, amendments to Title VII of the Civil Rights Act of 1964 to prohibit sex discrimination on the basis of pregnancy, October 31, 1978.

Public Law 96-511, Paperwork Reduction Act, December 11, 1980.

Public Law 101-336, Americans with Disabilities Act, July 26, 1990.

Public Law 103-3, Family and Medical Leave Act, February 5, 1993.

Public Law 103-353, Uniformed Services Employment and Reemployment Rights Act, October 13, 1994.

Public Law 104-275, Veterans' Benefits Improvement Act, October 9, 1996. As of March 20, 2013: http://www.gpo.gov/fdsys/pkg/PLAW-104publ275/pdf/PLAW-104publ275.pdf

Public Law 105-368, Veterans Programs Enhancement Act, November 11, 1998. As of March 20, 2013: http://www.gpo.gov/fdsys/pkg/PLAW-105publ368/pdf/PLAW-105publ368.pdf

Public Law 106-419, Veterans Benefits and Health Care Improvement Act, November 1, 2000. As of March 20, 2013: http://www.gpo.gov/fdsys/pkg/PLAW-106publ419/pdf/PLAW-106publ419.pdf

Public Law 108-454, Veterans Benefits Improvement Act, December 10, 2004. As of March 20, 2013: http://www.gpo.gov/fdsys/pkg/PLAW-108publ454/content-detail.html

Public Law 110-389, Veterans' Benefits Improvement Act, October 10, 2008. As of March 25, 2013: http://www.gpo.gov/fdsys/pkg/PLAW-110publ389/pdf/PLAW-110publ389.pdf

Public Law 112-56, Vow to Hire Heroes Act, November 21, 2011. As of March 25, 2013: http://www.gpo.gov/fdsys/pkg/PLAW-112publ56/pdf/PLAW-112publ56.pdf

Service Members Law Center, home page, undated (a). As of April 8, 2013: http://www.servicemembers-lawcenter.org/

———, Reserve Officers Association, home page, undated (b). As of January 10, 2013: http://www.roa.org/site/PageServer?pagename=Servicemembers_Law_Center

Sparks, Andrew P., "From the Desert to the Courtroom: The Uniformed Services Employment and Reemployment Rights Act," *Hastings Law Journal*, Vol. 61, No. 3, 2009–2010, pp. 773–800.

StataCorp, *Stata Statistical Software: Release 10*, College Station, Texas, 2007.

Stevens v. Tennessee Valley Authority, 687 F.2d 158, 6th Circuit, 1982.

Supporting Britain's Reservists and Employers, "Financial Assistance," undated, referenced August 1, 2012. As of March 26, 2013: http://www.sabre.mod.uk/Employers/The-Mobilisation-process/Financial-assistance.aspx

Under Secretary of Defense for Personnel and Readiness, *Civilian Employment and Reemployment Rights of Applicants for, and Service Members and Former Service Members of the Uniformed Services*, Washington, D.C., Department of Defense Instruction 1205.12, April 4, 1996, incorporating change 1, April 16, 1997. As of March 20, 2013: http://www.dtic.mil/whs/directives/corres/pdf/120512p.pdf

———, *Managing the Reserve Components as an Operational Force*, Washington, D.C., Department of Defense Directive 1200.17, October 29, 2008. As of March 20, 2013: http://www.dtic.mil/whs/directives/corres/pdf/120017p.pdf

———, *Reserve Components Common Personnel Data System (RCCPDS)*, Department of Defense Instruction 7730.54, May 20, 2011a. As of January 25, 2013: http://www.dtic.mil/whs/directives/corres/pdf/773054p.pdf

———, *Reserve Components Common Personnel Data System (RCCPDS): Reporting Procedures*, Washington, D.C.: Department of Defense Manual 7730.54-M, Vol. 1, May 25, 2011b. As of January 25, 2013: http://www.dtic.mil/whs/directives/corres/pdf/773054m_vol1.pdf

U.S. Code, Title 10, Armed Forces, Subtitle A, General Military Law, Part I, Organization and General Military Powers, Chapter 1, Definitions, Section 101, Definitions. As of March 20, 2013: http://www.gpo.gov/fdsys/granule/USCODE-2011-title10/USCODE-2011-title10-subtitleA-partI-chap1-sec101/content-detail.html

———, Title 38, Veterans' Benefits, Part III, Readjustment and Related Benefits, Chapter 43, Employment and Reemployment Rights of Members of the Uniformed Services. As of March 25, 2013: http://www.gpo.gov/fdsys/granule/USCODE-2011-title38/USCODE-2011-title38-partIII-chap43/content-detail.html

———, Title 42, The Public Health and Welfare, Chapter 126, Equal Opportunity for Individuals with Disabilities, Subchapter I, Employment, Section 12111, Definitions. As of March 25, 2013: http://www.gpo.gov/fdsys/pkg/USCODE-2011-title42/pdf/USCODE-2011-title42-chap126-subchapI-sec12111.pdf

U.S. Department of Defense, "Reserves Civilian Employment Information Program Announced," Press Release 240-04, March 31, 2004. As of January 25, 2013: http://www.defense.gov/Releases/Release.aspx?ReleaseID=7184

———, *Employer Support of the Guard and Reserve Annual Report*, Washington, D.C., 2009.

———, Status of Forces Survey of Reserve Component Members data, January 2011.

U.S. Department of Labor, *Performance and Accountability Report 2008*, November 17, 2008. As of March 25, 2013: http://www.dol.gov/_sec/media/reports/annual2008/

———, *Uniformed Services Employment and Reemployment Rights Act of 1994 (USERRA): Fiscal Year 2010 Annual Report to Congress*, Washington, D.C., July 2011. As of March 20, 2013: http://www.dol.gov/vets/programs/userra/FY2010%20USERRA%20Annual%20Report.pdf

U.S. Government Accountability Office, *Military Personnel: Additional Actions Needed to Improve Oversight of Reserve Employment Issues*, Washington, D.C., GAO-07-259, February 2007. As of March 20, 2013: http://purl.access.gpo.gov/GPO/LPS78822

Waterstone, Michael, "Returning Veterans and Disability Law," *Notre Dame Law Review*, Vol. 85, No. 3, March 2010, pp. 1081–1133.

Wedlund, Ryan, "Citizen Soldiers Fighting Terrorism: Reservists' Reemployment Rights," *William Mitchell Law Review*, Vol. 30, No. 3, 2004, pp. 797–844. As of March 20, 2013: http://www.wmitchell.edu/lawreview/Volume30/Issue3/1Wedlund.pdf

Werber, Laura, Margaret C. Harrell, Danielle M. Varda, Kimberly Curry Hall, Megan K. Beckett, and Stefanie Stern, *Deployment Experiences of Guard and Reserve Families: Implications for Support and Retention*, Santa Monica, Calif.: RAND Corporation, MG-645-OSD, 2008. As of March 20, 2013: http://www.rand.org/pubs/monographs/MG645.html